These safety symbols are used in laboratory and field investigations in this book to indicate possibl[e]
ing of each symbol and refer to this page often. *Remember to wash your hands thoroughly after completing lab procedures.*

PROTECTIVE EQUIPMENT Do not begin any lab without the proper protection equipment.

GOGGLES	Proper eye protection must be worn when performing or observing science activities that involve items or conditions as listed below.	**APRON** Wear an approved apron when using substances that could stain, wet, or destroy cloth.	**SOAP** Wash hands with soap and water before removing goggles and after all lab activities.	**GLOVES** Wear gloves when working with biological materials, chemicals, animals, or materials that can stain or irritate hands.

LABORATORY HAZARDS

Symbols	Potential Hazards	Precaution	Response
DISPOSAL	contamination of classroom or environment due to improper disposal of materials such as chemicals and live specimens	• DO NOT dispose of hazardous materials in the sink or trash can. • Dispose of wastes as directed by your teacher.	• If hazardous materials are disposed of improperly, notify your teacher immediately.
EXTREME TEMPERATURE	skin burns due to extremely hot or cold materials such as hot glass, liquids, or metals; liquid nitrogen; dry ice	• Use proper protective equipment, such as hot mitts and/or tongs, when handling objects with extreme temperatures.	• If injury occurs, notify your teacher immediately.
SHARP OBJECTS	punctures or cuts from sharp objects such as razor blades, pins, scalpels, and broken glass	• Handle glassware carefully to avoid breakage. • Walk with sharp objects pointed downward, away from you and others.	• If broken glass or injury occurs, notify your teacher immediately.
ELECTRICAL	electric shock or skin burn due to improper grounding, short circuits, liquid spills, or exposed wires	• Check condition of wires and apparatus for fraying or uninsulated wires, and broken or cracked equipment. • Use only GFCI-protected outlets	• DO NOT attempt to fix electrical problems. Notify your teacher immediately.
CHEMICAL	skin irritation or burns, breathing difficulty, and/or poisoning due to touching, swallowing, or inhalation of chemicals such as acids, bases, bleach, metal compounds, iodine, poinsettias, pollen, ammonia, acetone, nail polish remover, heated chemicals, mothballs, and any other chemicals labeled or known to be dangerous	• Wear proper protective equipment such as goggles, apron, and gloves when using chemicals. • Ensure proper room ventilation or use a fume hood when using materials that produce fumes. • NEVER smell fumes directly. • NEVER taste or eat any material in the laboratory.	• If contact occurs, immediately flush affected area with water and notify your teacher. • If a spill occurs, leave the area immediately and notify your teacher.
FLAMMABLE	unexpected fire due to liquids or gases that ignite easily such as rubbing alcohol	• Avoid open flames, sparks, or heat when flammable liquids are present.	• If a fire occurs, leave the area immediately and notify your teacher.
OPEN FLAME	burns or fire due to open flame from matches, Bunsen burners, or burning materials	• Tie back loose hair and clothing. • Keep flame away from all materials. • Follow teacher instructions when lighting and extinguishing flames. • Use proper protection, such as hot mitts or tongs, when handling hot objects.	• If a fire occurs, leave the area immediately and notify your teacher.
ANIMAL SAFETY	injury to or from laboratory animals	• Wear proper protective equipment such as gloves, apron, and goggles when working with animals. • Wash hands after handling animals.	• If injury occurs, notify your teacher immediately.
BIOLOGICAL	infection or adverse reaction due to contact with organisms such as bacteria, fungi, and biological materials such as blood, animal or plant materials	• Wear proper protective equipment such as gloves, goggles, and apron when working with biological materials. • Avoid skin contact with an organism or any part of the organism. • Wash hands after handling organisms.	• If contact occurs, wash the affected area and notify your teacher immediately.
FUME	breathing difficulties from inhalation of fumes from substances such as ammonia, acetone, nail polish remover, heated chemicals, and mothballs	• Wear goggles, apron, and gloves. • Ensure proper room ventilation or use a fume hood when using substances that produce fumes. • NEVER smell fumes directly.	• If a spill occurs, leave area and notify your teacher immediately.
IRRITANT	irritation of skin, mucous membranes, or respiratory tract due to materials such as acids, bases, bleach, pollen, mothballs, steel wool, and potassium permanganate	• Wear goggles, apron, and gloves. • Wear a dust mask to protect against fine particles.	• If skin contact occurs, immediately flush the affected area with water and notify your teacher.
RADIOACTIVE	excessive exposure from alpha, beta, and gamma particles	• Remove gloves and wash hands with soap and water before removing remainder of protective equipment.	• If cracks or holes are found in the container, notify your teacher immediately.

Your online portal to everything you need

connectED.mcgraw-hill.com

Look for these icons to access exciting digital resources

 Video

 Audio

Review

Inquiry

WebQuest

Assessment

Concepts in Motion

GEOLOGIC CHANGES

iSCIENCE

Glencoe

Geode

This is a cross section of geode, a type of rock. The outside of a geode is generally limestone, but the inside contains mineral crystals. Crystals only partially fill this geode, but other geodes are filled completely with crystals.

The McGraw·Hill Companies

 Education

Send all inquiries to:
McGraw-Hill Education
8787 Orion Place
Columbus, OH 43240-4027

ISBN: 978-0-07-888009-4
MHID: 0-07-888009-2

Printed in the United States of America.

2 3 4 5 6 7 8 9 10 DOW 15 14 13 12 11

Authors and Contributors

Authors

American Museum of Natural History
New York, NY

Michelle Anderson, MS
Lecturer
The Ohio State University
Columbus, OH

Juli Berwald, PhD
Science Writer
Austin, TX

John F. Bolzan, PhD
Science Writer
Columbus, OH

Rachel Clark, MS
Science Writer
Moscow, ID

Patricia Craig, MS
Science Writer
Bozeman, MT

Randall Frost, PhD
Science Writer
Pleasanton, CA

Lisa S. Gardiner, PhD
Science Writer
Denver, CO

Jennifer Gonya, PhD
The Ohio State University
Columbus, OH

Mary Ann Grobbel, MD
Science Writer
Grand Rapids, MI

Whitney Crispen Hagins, MA, MAT
Biology Teacher
Lexington High School
Lexington, MA

Carole Holmberg, BS
Planetarium Director
Calusa Nature Center and
Planetarium, Inc.
Fort Myers, FL

Tina C. Hopper
Science Writer
Rockwall, TX

Jonathan D. W. Kahl, PhD
Professor of Atmospheric Science
University of Wisconsin-
Milwaukee
Milwaukee, WI

Nanette Kalis
Science Writer
Athens, OH

S. Page Keeley, MEd
Maine Mathematics and
Science Alliance
Augusta, ME

Cindy Klevickis, PhD
Professor of Integrated Science
and Technology
James Madison University
Harrisonburg, VA

Kimberly Fekany Lee, PhD
Science Writer
La Grange, IL

Michael Manga, PhD
Professor
University of California, Berkeley
Berkeley, CA

Devi Ried Mathieu
Science Writer
Sebastopol, CA

Elizabeth A. Nagy-Shadman, PhD
Geology Professor
Pasadena City College
Pasadena, CA

William D. Rogers, DA
Professor of Biology
Ball State University
Muncie, IN

Donna L. Ross, PhD
Associate Professor
San Diego State University
San Diego, CA

Marion B. Sewer, PhD
Assistant Professor
School of Biology
Georgia Institute of Technology
Atlanta, GA

Julia Meyer Sheets, PhD
Lecturer
School of Earth Sciences
The Ohio State University
Columbus, OH

Michael J. Singer, PhD
Professor of Soil Science
Department of Land, Air and
Water Resources
University of California
Davis, CA

Karen S. Sottosanti, MA
Science Writer
Pickerington, Ohio

Paul K. Strode, PhD
I.B. Biology Teacher
Fairview High School
Boulder, CO

Jan M. Vermilye, PhD
Research Geologist
Seismo-Tectonic Reservoir
Monitoring (STRM)
Boulder, CO

Judith A. Yero, MA
Director
Teacher's Mind Resources
Hamilton, MT

Dinah Zike, MEd
Author, Consultant,
Inventor of Foldables
Dinah Zike Academy;
Dinah-Might Adventures, LP
San Antonio, TX

Margaret Zorn, MS
Science Writer
Yorktown, VA

Consulting Authors

Alton L. Biggs
Biggs Educational Consulting
Commerce, TX

Ralph M. Feather, Jr., PhD
Assistant Professor
Department of Educational
Studies and Secondary
Education
Bloomsburg University
Bloomsburg, PA

Douglas Fisher, PhD
Professor of Teacher Education
San Diego State University
San Diego, CA

Edward P. Ortleb
Science/Safety Consultant
St. Louis, MO

Series Consultants

Science

Solomon Bililign, PhD
Professor
Department of Physics
North Carolina Agricultural
and Technical State University
Greensboro, NC

John Choinski
Professor
Department of Biology
University of Central Arkansas
Conway, AR

Anastasia Chopelas, PhD
Research Professor
Department of Earth and
Space Sciences
UCLA
Los Angeles, CA

David T. Crowther, PhD
Professor of Science Education
University of Nevada, Reno
Reno, NV

A. John Gatz
Professor of Zoology
Ohio Wesleyan University
Delaware, OH

Sarah Gille, PhD
Professor
University of California
San Diego
La Jolla, CA

David G. Haase, PhD
Professor of Physics
North Carolina State
University
Raleigh, NC

Janet S. Herman, PhD
Professor
Department of Environmental
Sciences
University of Virginia
Charlottesville, VA

David T. Ho, PhD
Associate Professor
Department of Oceanography
University of Hawaii
Honolulu, HI

Ruth Howes, PhD
Professor of Physics
Marquette University
Milwaukee, WI

**Jose Miguel Hurtado, Jr.,
PhD**
Associate Professor
Department of Geological
Sciences
University of Texas at El Paso
El Paso, TX

Monika Kress, PhD
Assistant Professor
San Jose State University
San Jose, CA

Mark E. Lee, PhD
Associate Chair & Assistant
Professor
Department of Biology
Spelman College
Atlanta, GA

Linda Lundgren
Science writer
Lakewood, CO

Series Consultants, continued

Keith O. Mann, PhD
Ohio Wesleyan University
Delaware, OH

Charles W. McLaughlin, PhD
Adjunct Professor of
Chemistry
Montana State University
Bozeman, MT

Katharina Pahnke, PhD
Research Professor
Department of Geology and
Geophysics
University of Hawaii
Honolulu, HI

Jesús Pando, PhD
Associate Professor
DePaul University
Chicago, IL

Hay-Oak Park, PhD
Associate Professor
Department of Molecular
Genetics
Ohio State University
Columbus, OH

David A. Rubin, PhD
Associate Professor of
Physiology
School of Biological Sciences
Illinois State University
Normal, IL

Toni D. Sauncy
Assistant Professor of Physics
Department of Physics
Angelo State University
San Angelo, TX

Malathi Srivatsan, PhD
Associate Professor
of Neurobiology
College of Sciences and
Mathematics
Arkansas State University
Jonesboro, AR

Cheryl Wistrom, PhD
Associate Professor
of Chemistry
Saint Joseph's College
Rensselaer, IN

Reading

ReLeah Cossett Lent
Author/Educational
Consultant
Blue Ridge, GA

Math

Vik Hovsepian
Professor of Mathematics
Rio Hondo College
Whittier, CA

Series Reviewers

Thad Boggs
Mandarin High School
Jacksonville, FL

Catherine Butcher
Webster Junior High School
Minden, LA

Erin Darichuk
West Frederick Middle School
Frederick, MD

Joanne Hedrick Davis
Murphy High School
Murphy, NC

Anthony J. DiSipio, Jr.
Octorara Middle School
Atglen, PA

Adrienne Elder
Tulsa Public Schools
Tulsa, OK

Series Reviewers, continued

Carolyn Elliott
Iredell-Statesville Schools
Statesville, NC

Christine M. Jacobs
Ranger Middle School
Murphy, NC

Jason O. L. Johnson
Thurmont Middle School
Thurmont, MD

Felecia Joiner
Stony Point Ninth Grade Center
Round Rock, TX

Joseph L. Kowalski, MS
Lamar Academy
McAllen, TX

Brian McClain
Amos P. Godby High School
Tallahassee, FL

Von W. Mosser
Thurmont Middle School
Thurmont, MD

Ashlea Peterson
Heritage Intermediate Grade Center
Coweta, OK

Nicole Lenihan Rhoades
Walkersville Middle School
Walkersvillle, MD

Maria A. Rozenberg
Indian Ridge Middle School
Davie, FL

Barb Seymour
Westridge Middle School
Overland Park, KS

Ginger Shirley
Our Lady of Providence Junior-Senior High School
Clarksville, IN

Curtis Smith
Elmwood Middle School
Rogers, AR

Sheila Smith
Jackson Public School
Jackson, MS

Sabra Soileau
Moss Bluff Middle School
Lake Charles, LA

Tony Spoores
Switzerland County Middle School
Vevay, IN

Nancy A. Stearns
Switzerland County Middle School
Vevay, IN

Kari Vogel
Princeton Middle School
Princeton, MN

Alison Welch
Wm. D. Slider Middle School
El Paso, TX

Linda Workman
Parkway Northeast Middle School
Creve Coeur, MO

Teacher Advisory Board

Online Guide

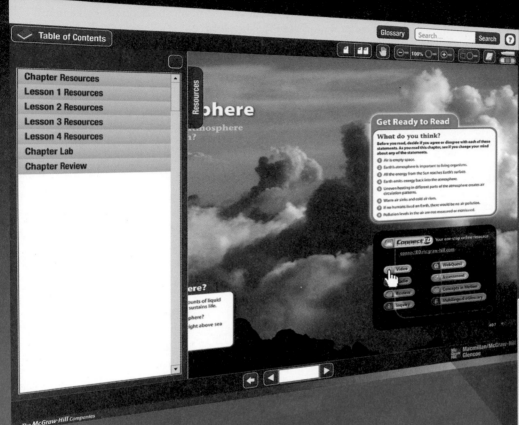

Video

ConnectED

▷ **Your Digital Science Portal**

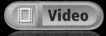 **Video**

See the science in real life through these exciting videos.

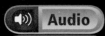 **Audio**

Click the link and you can listen to the text while you follow along.

 **Review**

Try these interactive tools to help you review the lesson concepts.

 **Inquiry**

Explore concepts through hands–on and virtual labs.

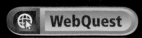 **WebQuest**

These web-based challenges relate the concepts you're learning about to the latest news and research.

The icons in your online student edition link you to interactive learning opportunities. Browse your online student book to find more.

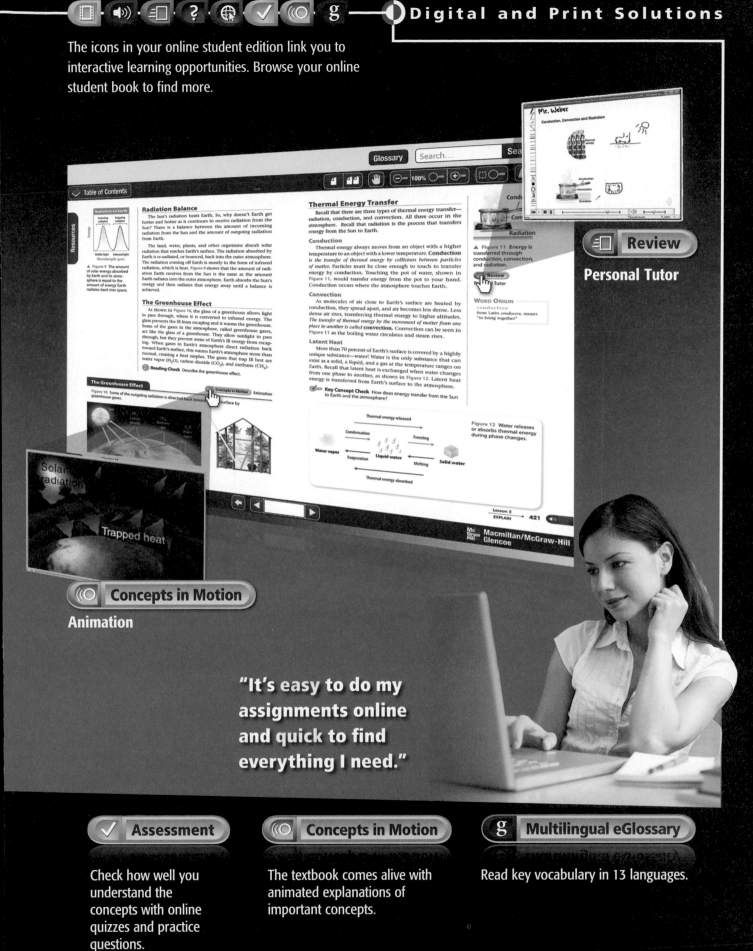

Concepts in Motion

Animation

Review

Personal Tutor

"It's easy to do my assignments online and quick to find everything I need."

✓ **Assessment**

Check how well you understand the concepts with online quizzes and practice questions.

◉ **Concepts in Motion**

The textbook comes alive with animated explanations of important concepts.

g **Multilingual eGlossary**

Read key vocabulary in 13 languages.

Treasure Hunt

Your science book has many features that will aid you in your learning. Some of these features are listed below. You can use the activity at the right to help you find these and other special features in the book.

- **THE BIG IDEA** can be found at the start of each chapter.

- The Reading Guide at the start of each lesson lists 🔑 **Key Concepts**, vocabulary terms, and online supplements to the content.

- 🖥️ **Connect ED** icons direct you to online resources such as animations, personal tutors, math practices, and quizzes.

- **Inquiry** Labs and Skill Practices are in each chapter.

- Your **FOLDABLES** help organize your notes.

1 What four margin items can help you build your vocabulary?

2 On what page does the glossary begin? What glossary is online?

3 In which Student Resource at the back of your book can you find a listing of Laboratory Safety Symbols?

4 Suppose you want to find a list of all the Launch Labs, MiniLabs, Skill Practices, and Labs, where do you look?

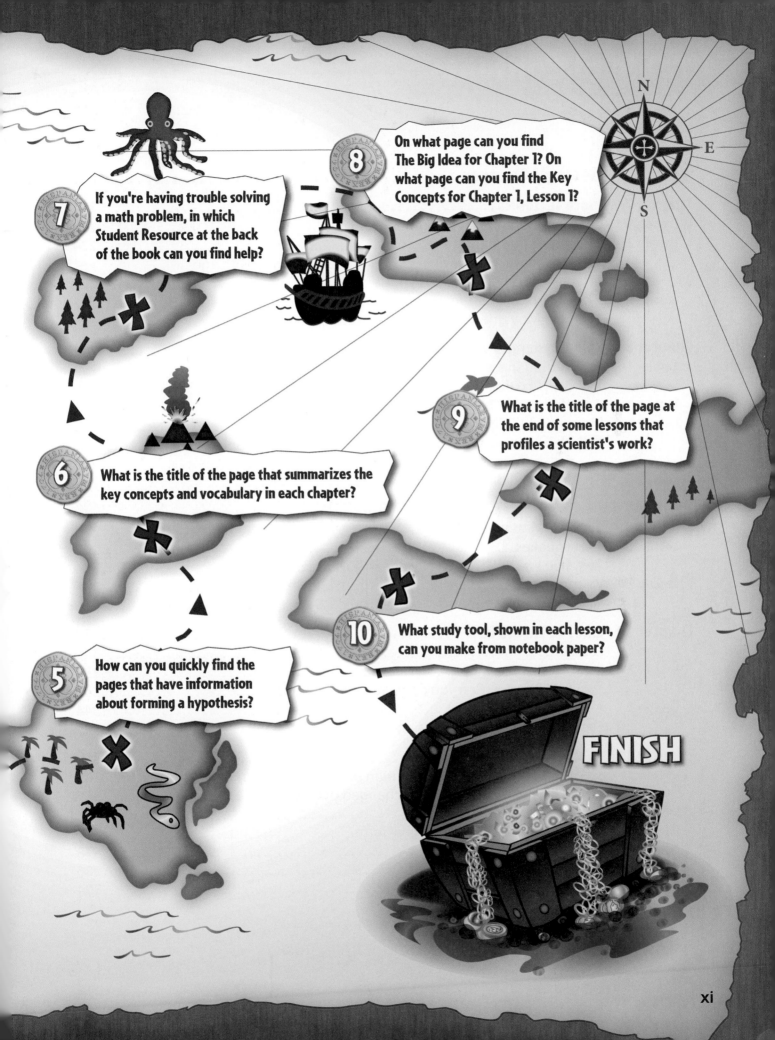

7 If you're having trouble solving a math problem, in which Student Resource at the back of the book can you find help?

8 On what page can you find The Big Idea for Chapter 1? On what page can you find the Key Concepts for Chapter 1, Lesson 1?

9 What is the title of the page at the end of some lessons that profiles a scientist's work?

6 What is the title of the page that summarizes the key concepts and vocabulary in each chapter?

10 What study tool, shown in each lesson, can you make from notebook paper?

5 How can you quickly find the pages that have information about forming a hypothesis?

FINISH

Table of Contents

TABLE OF CONTENTS

Student Resources

TABLE OF CONTENTS

Inquiry

Inquiry Launch Labs

Inquiry MiniLabs

Inquiry Skill Practice

Inquiry Labs

Features

HOW NATURE WORKS

CAREERS in SCIENCE

TABLE OF CONTENTS

Unit 2

Geologic Changes

5 Billion B.C.   1700   1800

4.57 billion years ago
The Sun forms.

4.54 billion years ago
Earth forms.

1778
French naturalist Comte du Buffon creates a small globe resembling Earth and measures its cooling rate to estimate Earth's age. He concludes that Earth is approximately 75,000 years old.

1830
Geologist Charles Lyell begins publishing *The Principles of Geology*; his work popularizes the concept that the features of Earth are perpetually changing, eroding, and reforming.

1862
Physicist William Thomson publishes calculations that Earth is approximately 20 million years old. He claims that Earth had formed as a completely molten object, and he calculates the amount of time it would take for the surface to cool to its present temperature.

1900   2000

1899–1900
John Joly releases his findings from calculating the age of Earth using the rate of oceanic salt accumulation. He determines that the oceans are about 80–100 million years old.

1905
Ernest Rutherford and Bertrand Boltwood use radiometric dating to determine the age of rock samples. This technique would later be used to determine the age of Earth.

1956
Today's accepted age of Earth is determined by C.C. Patterson using uranium-lead isotope dating on several meteorites.

? Inquiry
Visit ConnectED for this unit's **STEM** activity.

Science and History

About 500,000 years ago, early humans used stone to make tools, weapons, and small decorative items. Then, about 8,000 years ago, someone might have spied a shiny object among the rocks. It was gold—thought to be the first metal discovered by humans. Gold was very different from stone. It did not break when it was struck. It could easily be shaped into useful and beautiful objects. Over time, other metals were discovered. Each metal helped advance human civilization. Metals from Earth's crust have helped humans progress from the Stone Age to the Moon, to Mars, and beyond.

Gold

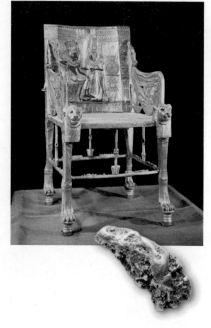

Since the time of its discovery, gold has been a symbol of wealth and power. It is used mainly in jewelry, coins, and other valuable objects. King Tut's coffin was made of pure gold. Tut's body was surrounded by the largest collection of gold objects ever discovered—chariots, statues, jewelry, and a golden throne. Because gold is so valuable, much of it is recycled. If you own a piece of gold jewelry, it might contain gold that was mined thousands of years ago!

Lead

Ancient Egyptians used the mineral lead sulfide, also called galena, as eye paint. About 5,500 years ago, metalworkers found that galena melts at a low temperature, forming puddles of the lead. Lead bends easily, and the Romans shaped it into pipes for carrying water. Over the years, the Romans realized that lead was entering the water and was toxic to humans. Despite possible danger, lead water pipes were common in modern homes for decades. Finally, however, in 2004 the use of lead pipes in home construction was banned.

Copper

The first metal commonly traded was copper. About 5,000 years ago, Native Americans mined more than half a million tons of copper from the area that is now Michigan. Copper is stronger than gold. Back then, it was shaped into saws, axes, and other tools. Stronger saws made it easier to cut down trees. The wood from trees then could be used to build boats, which allowed trade routes to expand. Many cultures today still use methods to shape copper that are similar to those used by ancient peoples.

Tin and Bronze

Around 4,500 years ago, the Sumerians noticed differences in the copper they used. Some flowed more easily when it melted and was stronger after it hardened. They discovered that this harder copper contained another metal—tin. Metalworkers began combining tin and copper to produce a metal called bronze. Bronze eventually replaced copper as the most important metal to society. Bronze was strong and cheap enough to make everyday tools. It could easily be shaped into arrowheads, armor, axes, and sword blades. People admired the appearance of bronze. It continues to be used in sculptures. Bronze, along with gold and silver, is used in Olympic medals as a symbol of excellence.

Iron and Steel

Although iron-containing rock was known centuries ago, people couldn't build fires hot enough to melt the rock and separate out the iron. As fire-building methods improved, iron use became more common. It replaced bronze for all uses except art. Iron farm tools revolutionized agriculture. Iron weapons became the choice for war. Like metals used by earlier civilizations, iron increased trade and wealth, and improved people's lives.

In the 17th century, metalworkers developed a way to mix iron with carbon. This process formed steel. Steel quickly became valued for its strength, resistance to rusting, and ease of use in welding. Besides being used in the construction of skyscrapers, bridges, and highways, steel is used to make tools, ships, vehicles, machines, and appliances.

Try to imagine your world without metals. Throughout history, metals changed society as people learned to use them.

Inquiry MiniLab
20 minutes

How do a metal's properties affect its uses?

Why are different common objects made of a variety of different metals?

1. Read and complete a lab safety form.

2. Examine a **lead fishing weight,** a piece of **copper tubing,** and an **iron bolt.**

3. Create a table comparing characteristics of the objects in your Science Journal.

4. Use a **hammer** to tap on each item. Record your observations in your table.

Analyze and Conclude

1. **Infer** Why was lead, not copper or iron, used to make the fishing weight?

2. **Compare** What similarities do all three objects share?

3. **Infer** Why do you think ancient peoples used lead for pipes and iron for weapons?

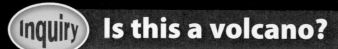

Chapter 7

Plate Tectonics

THE BIG IDEA What is the theory of plate tectonics?

Inquiry Is this a volcano?

Iceland is home to many active volcanoes like this one. This eruption is called a fissure eruption. This occurs when lava erupts from a long crack, or fissure, in Earth's crust.

- Why is the crust breaking apart here?

- What factors determine where a volcano will form?

- How are volcanoes associated with plate tectonics?

Get Ready to Read

What do you think?

Before you read, decide if you agree or disagree with each of these statements. As you read this chapter, see if you change your mind about any of the statements.

1 India has always been north of the equator.

2 All the continents once formed one supercontinent.

3 The seafloor is flat.

4 Volcanic activity occurs only on the seafloor.

5 Continents drift across a molten mantle.

6 Mountain ranges can form when continents collide.

 **Connect ED** Your one-stop online resource

connectED.mcgraw-hill.com

 Video

 WebQuest

 Audio

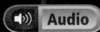 Assessment

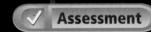 Review

Concepts in Motion

 Inquiry

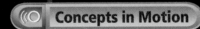 Multilingual eGlossary

Lesson 1

Reading Guide

Key Concepts
ESSENTIAL QUESTIONS

- What evidence supports continental drift?
- Why did scientists question the continental drift hypothesis?

Vocabulary

Pangaea p. 217

continental drift p. 217

 g **Multilingual eGlossary**

Video **BrainPOP®**

The Continental Drift Hypothesis

Inquiry — How did this happen?

In Iceland, elongated cracks called rift zones are easy to find. Why do rift zones occur here? Iceland is above an area of the seafloor where Earth's crust is breaking apart. Earth's crust is constantly on the move. Scientists realized this long ago, but they could not prove how or why this happened.

🔊 **216** Chapter 7
ENGAGE

Can you put together a peel puzzle?

Early map makers observed that the coastlines of Africa and South America appeared as if they could fit together like pieces of a puzzle. Scientists eventually discovered that these continents were once part of a large landmass. Can you use an orange peel to illustrate how continents may have fit together?

1 Read and complete a lab safety form.

2 Carefully peel an **orange,** keeping the orange-peel pieces as large as possible.

3 Set the orange aside.

4 Refit the orange-peel pieces back together in the shape of a sphere.

5 After successfully reconstructing the orange peel, disassemble your pieces.

6 Trade the entire orange peel with a classmate and try to reconstruct his or her orange peel.

Think About This

1. Which orange peel was easier for you to reconstruct? Why?

2. Look at a world map. Do the coastlines of any other continents appear to fit together?

3. **Key Concept** What additional evidence would you need to prove that all the continents might have once fit together?

Pangaea

Did you know that Earth's surface is on the move? Can you feel it? Each year, North America moves a few centimeters farther away from Europe and closer to Asia. That is several centimeters, or about the thickness of this book. Even though you don't necessarily feel this motion, Earth's surface moves slowly every day.

Nearly 100 years ago Alfred Wegener (VAY guh nuhr), a German scientist, began an important investigation that continues today. Wegener wanted to know whether Earth's continents were fixed in their positions. He proposed that *all the continents were once part of a supercontinent called* **Pangaea** (pan JEE uh). Over time Pangaea began breaking apart, and the continents slowly moved to their present positions. Wegener proposed the hypothesis of **continental drift,** *which suggested that continents are in constant motion on the surface of Earth.*

Alfred Wegener observed the similarities of continental coastlines now separated by oceans. Look at the outlines of Africa and South America in **Figure 1.** Notice how they could fit together like pieces of a puzzle. Hundreds of years ago mapmakers noticed this jigsaw-puzzle pattern as they made the first maps of the continents.

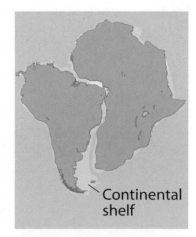
Continental shelf

Figure 1 The eastern coast of South America mirrors the shape of the west coast of Africa.

Evidence That Continents Move

If you had discovered continental drift, how would you have tested your hypothesis? The most obvious evidence for continental drift is that the continents appear to fit together like pieces of a puzzle. But scientists were skeptical, and Wegener needed additional evidence to help support his hypothesis.

Climate Clues

When Wegener pieced Pangaea together, he proposed that South America, Africa, India, and Australia were located closer to Antarctica 280 million years ago. He suggested that the climate of the Southern Hemisphere was much cooler at the time. Glaciers covered large areas that are now parts of these continents. These glaciers would have been similar to the ice sheet that covers much of Antarctica today.

Wegener used climate clues to support his continental drift hypothesis. He studied the sediments deposited by glaciers in South America and Africa, as well as in India and Australia. Beneath these sediments, Wegener discovered glacial grooves, or deep scratches in rocks made as the glaciers moved across land. **Figure 2** shows where these glacial features are found on neighboring continents today. These continents were once part of the supercontinent Pangaea, when the climate in the Southern Hemisphere was cooler.

Climate Clues 🔑

Concepts in Motion **Animation**

Figure 2 If the southern hemisphere continents could be reassembled into Pangaea, the presence of an ice sheet would explain the glacial features on these continents today.

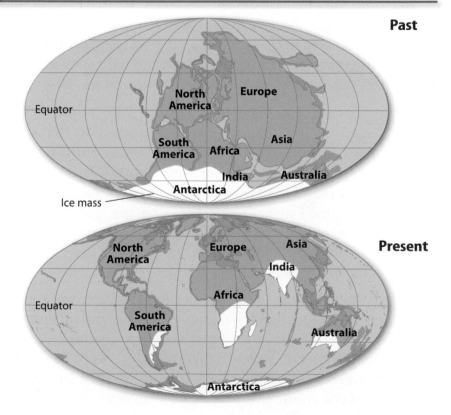

Fossil Clues

Animals and plants that live on different continents can be unique to that continent alone. Lions live in Africa but not in South America. Kangaroos live in Australia but not on any other continent. Because oceans separate continents, these animals cannot travel from one continent to another by natural means. However, fossils of similar organisms have been found on several continents separated by oceans. How did this happen? Wegener argued that these continents must have been connected some time in the past.

Fossils of a plant called *Glossopteris* (glahs AHP tur us) have been discovered in rocks from South America, Africa, India, Australia, and Antarctica. These continents are far apart today. The plant's seeds could not have traveled across the vast oceans that separate them. **Figure 3** shows that when these continents were part of Pangaea 225 million years ago, *Glossopteris* lived in one region. Evidence suggests these plants grew in a swampy environment. Therefore, the climate of this region, including Antarctica, was different than it is today. Antarctica had a warm and wet climate. The climate had changed drastically from what it was 55 million years earlier when glaciers existed.

Reading Check How did climate in Antarctica change between 280 and 225 million years ago?

Fossil Clues ⚷

Figure 3 Fossils of *Glossopteris* have been found on many continents that are now separated by oceans. The orange area in the image on the right represents where *Glossopteris* fossils have been found.

Visual Check Which of the continents would not support *Glossopteris* growth today?

> **REVIEW VOCABULARY**
>
> fossil
> the naturally preserved remains, imprints, or traces of organisms that lived long ago

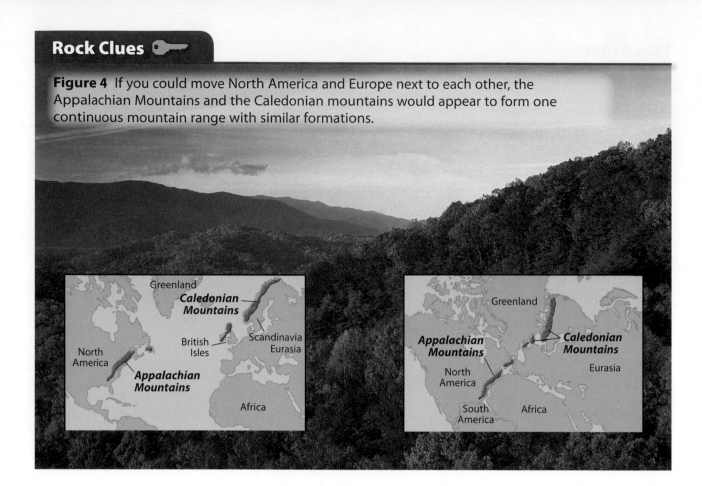

Figure 4 If you could move North America and Europe next to each other, the Appalachian Mountains and the Caledonian mountains would appear to form one continuous mountain range with similar formations.

FOLDABLES®

Make a horizontal half-book and write the title as shown. Use it to organize your notes on the continental drift hypothesis.

Evidence for the Continental Drift Hypothesis

Rock Clues

Wegener realized he needed more evidence to support the continental drift hypothesis. He observed that mountain ranges like the ones shown in **Figure 4** and rock formations on different continents had common origins. Today, geologists can determine when these rocks formed. For example, geologists suggest that large-scale volcanic eruptions occurred on the western coast of Africa and the eastern coast of South America at about the same time hundreds of millions of years ago. The volcanic rocks from the eruptions are identical in both chemistry and age. Refer back to **Figure 1.** If you could superimpose similar rock types onto the maps, these rocks would be in the area where Africa and South America fit together.

The Caledonian mountain range in northern Europe and the Appalachian Mountains in eastern North America are similar in age and structure. They are also composed of the same rock types. If you placed North America and Europe next to each other, these mountains would meet and form one long, continuous mountain belt. **Figure 4** illustrates where this mountain range would be.

✓ **Key Concept Check** How were similar rock types used to support the continental drift hypothesis?

What was missing?

Wegener continued to support the continental drift hypothesis until his death in 1930. Wegener's ideas were not widely accepted until nearly four decades later. Why were scientists skeptical of Wegener's hypothesis? Although Wegener had evidence to suggest that continents were on the move, he could not explain how they moved.

One reason scientists questioned continental drift was because it is a slow process. It was not possible for Wegener to measure how fast the continents moved. The main objection to the continental drift hypothesis, however, was that Wegener could not explain what forces caused the continents to move. The mantle beneath the continents and the seafloor is made of solid rock. How could continents push their way through solid rock? Wegener needed more scientific evidence to prove his hypothesis. However, this evidence was hidden on the seafloor between the drifting continents. The evidence necessary to prove continental drift was not discovered until long after Wegener's death.

 Key Concept Check Why did scientists argue against Wegener's continental drift hypothesis?

SCIENCE USE v. COMMON USE

mantle

Science Use the middle layer of Earth, situated between the crust above and the core below

Common Use a loose, sleeveless garment worn over other clothes

---

Inquiry MiniLab   **20 minutes**

How do you use clues to put puzzle pieces together?

When you put a puzzle together, you use clues to figure out which pieces fit next to each other. How did Wegener use a similar technique to piece together Pangaea?

1 Read and complete a lab safety form.

2 Using **scissors,** cut a piece of **newspaper** or a page from a **magazine** into an irregular shape with a diameter of about 25 cm.

3 Cut the piece of paper into at least 12 but not more than 20 pieces.

4 Exchange your puzzle with a partner and try to fit the new puzzle pieces together.

5 Reclaim your puzzle and remove any three pieces. Exchange your incomplete puzzle with a different partner. Try to put the incomplete puzzles back together.

Analyze and Conclude

1. **Summarize** Make a list of the clues you used to put together your partner's puzzle.

2. **Describe** How was putting together a complete puzzle different from putting together an incomplete puzzle?

3. **Key Concept** What clues did Wegener use to hypothesize the existence of Pangaea? What clues were missing from Wegener's puzzle?

Lesson 1 Review

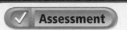
Visual Summary

Past

All continents were once part of a super-continent called Pangaea.

Present

Evidence found on present-day continents suggests that the continents have moved across Earth's surface.

FOLDABLES

Use your lesson Foldable to review the lesson. Save your Foldable for the project at the end of the chapter.

What do you think NOW?

You first read the statements below at the beginning of the chapter.

1. India has always been north of the equator.

2. All the continents once formed one supercontinent.

Did you change your mind about whether you agree or disagree with the statements? Rewrite any false statements to make them true.

Use Vocabulary

1 **Define** *Pangaea*.

2 **Explain** the continental drift hypothesis and the evidence used to support it.

Understand Key Concepts

3 **Identify** the scientist who first proposed that the continents move away from or toward each other.

4 Which can be used as an indicator of past climate?
- **A.** fossils
- **B.** lava flows
- **C.** mountain ranges
- **D.** tides

Interpret Graphics

5 **Interpret** Look at the map of the continents below. What direction has South America moved relative to Africa?

6 **Summarize** Copy and fill in the graphic organizer below to show the evidence Alfred Wegener used to support his continental drift hypothesis.

Continental Drift Hypothesis

Critical Thinking

7 **Recognize** The shape and age of the Appalachian Mountains are similar to the Caledonian mountains in northern Europe. What else could be similar?

8 **Explain** If continents continue to drift, is it possible that a new supercontinent will form? Which continents might be next to each other 200 million years from now?

▼ This small mammal is a close living relative of an animal that once roamed Antarctica.

Gondwana

▲ Ross MacPhee is a paleontologist working for the American Museum of Natural History in New York City. Here, he is searching for fossils in Antarctica.

A Fossil Clue from the Giant Landmass that Once Dominated the Southern Hemisphere

If you could travel back in time 120 million years, you would probably discover that Earth looked very different than it does today. Scientists believe that instead of seven continents, there were two giant landmasses, or supercontinents, on Earth at that time. Scientists named the landmass in the northern hemisphere *Laurasia*. The landmass in the southern hemisphere was named *Gondwana*. It included the present-day continents of Antarctica, South America, Australia, and Africa.

How do scientists know that Gondwana existed? Ross MacPhee is a paleontologist—a scientist who studies fossils. MacPhee recently traveled to Antarctica where he discovered the fossilized tooth of a small land mammal. After carefully examining the tooth, he realized that it resembled fossils from ancient land mammals found in Africa and North America. MacPhee believes that these mammals are the ancient relatives of a mammal living today on the African island-nation of Madagascar.

How did the fossil remains and their present-day relatives become separated by kilometers of ocean? MacPhee hypothesizes that the mammal migrated across land bridges that once connected parts of Gondwana. Over millions of years, the movement of Earth's tectonic plates broke up this supercontinent. New ocean basins formed between the continents, resulting in the arrangement of landmasses that we see today.

LAURASIA
North America
Europe and Asia

GONDWANA
Africa
South America
Arabia
India
Australia
Antarctica

▲ *Gondwana* and *Laurasia* formed as the supercontinent Pangaea broke apart.

It's Your Turn

RESEARCH Millions of years ago, the island of Madagascar separated from the continent of Gondwana. In this environment, the animals of Madagascar changed and adapted. Research and report on one animal. Describe some of its unique adaptations.

Reading Guide

Key Concepts 🔑
ESSENTIAL QUESTIONS

- What is seafloor spreading?
- What evidence is used to support seafloor spreading?

Vocabulary
mid-ocean ridge p. 225
seafloor spreading p. 226
normal polarity p. 228
magnetic reversal p. 228
reversed polarity p. 228

g Multilingual eGlossary

Development of a Theory

Inquiry What do the colors represent?

The colors in this satellite image show topography. The warm colors, red, pink, and yellow, represent landforms above sea level. The greens and blues indicate changes in topography below sea level. Deep in the Atlantic Ocean there is a mountain range, shown here as a linear feature in green. Is there a connection between this landform and the continental drift hypothesis?

Can you guess the age of the glue?

The age of the seafloor can be determined by measuring magnetic patterns in rocks from the bottom of the ocean. How can similar patterns in drying glue be used to show age relationships between rocks exposed on the seafloor?

1 Read and complete a lab safety form.

2 Carefully spread a thin layer of **rubber cement** on a sheet of **paper.**

3 Observe for 3 minutes. Record the pattern of how the glue dries in your Science Journal.

4 Repeat step 2. After 1 minute, exchange papers with a classmate.

5 Ask the classmate to observe and tell you which part of the glue dried first.

Think About This

1. What evidence helped you to determine the oldest and youngest glue layers?

2. How is this similar to a geologist trying to estimate the age of rocks on the seafloor?

3. 🔑 **Key Concept** How could magnetic patterns in rock help predict a rock's age?

Mapping the Ocean Floor

During the late 1940s after World War II, scientists began exploring the seafloor in greater detail. They were able to determine the depth of the ocean using a device called an echo sounder, as shown in **Figure 5.** Once ocean depths were determined, scientists used these data to create a topographic map of the seafloor. These new topographic maps of the seafloor revealed that vast mountain ranges stretched for many miles deep below the ocean's surface. *The mountain ranges in the middle of the oceans are called* **mid-ocean ridges.** Mid-ocean ridges, shown in **Figure 5,** are much longer than any mountain range on land.

Figure 5 An echo sounder produces sound waves that travel from a ship to the seafloor and back. The deeper the ocean, the longer the time this takes. Depth can be used to determine seafloor topography.

Seafloor Topography

Mid-ocean Ridge

Sediment

Magma

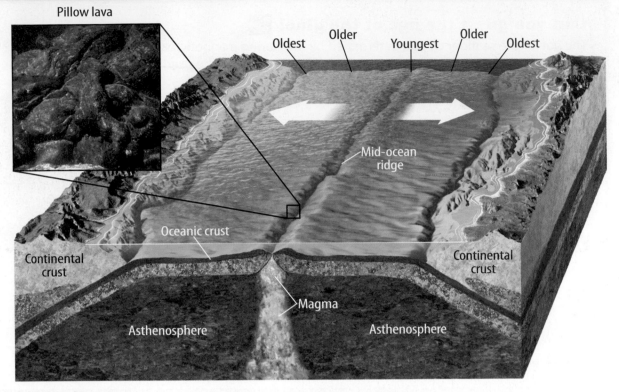

Pillow lava

Oldest   Older   Youngest   Older   Oldest

Mid-ocean ridge

Oceanic crust

Continental crust   Continental crust

Magma

Asthenosphere   Asthenosphere

Figure 6 When lava erupts along a mid-ocean ridge, it cools and crystallizes, forming a type of rock called basalt. Basalt is the dominant rock on the seafloor. The youngest basalt is closest to the ridge. The oldest basalt is farther away from the ridge.

✓ **Visual Check** Looking at the image above, can you propose a pattern that exists in rocks on either side of the mid-ocean ridge?

Seafloor Spreading

By the 1960s scientists discovered a new process that helped explain continental drift. This process, shown in **Figure 6,** is called seafloor spreading. **Seafloor spreading** *is the process by which new oceanic crust forms along a mid-ocean ridge and older oceanic crust moves away from the ridge.*

When the seafloor spreads, the mantle below melts and forms magma. Because magma is less dense than solid mantle material, it rises through cracks in the crust along the mid-ocean ridge. When magma erupts on Earth's surface, it is called lava. As this lava cools and crystallizes on the seafloor, it forms a type of rock called basalt. Because the lava erupts into water, it cools rapidly and forms rounded structures called pillow lavas. Notice the shape of the pillow lava shown in **Figure 6.**

As the seafloor continues to spread apart, the older oceanic crust moves away from the mid-ocean ridge. The closer the crust is to a mid-ocean ridge, the younger the oceanic crust is. Scientists argued that if the seafloor spreads, the continents must also be moving. A mechanism to explain continental drift was finally discovered long after Wegener proposed his hypothesis.

🔑 **Key Concept Check** What is seafloor spreading?

Topography of the Seafloor

The rugged mountains that make up the mid-ocean ridge system can form in two different ways. For example, large amounts of lava can erupt from the center of the ridge, cool and build up around the ridge. Or, as the lava cools and forms new crust, it cracks. The rocks move up or down along these cracks in the seafloor, forming jagged mountain ranges.

Reading Check How do mountains form along the mid-ocean ridge?

Over time, sediment accumulates on top of the oceanic crust. Close to the mid-ocean ridge there is almost no sediment. Far from the mid-ocean ridge, the layer of sediment becomes thick enough to make the seafloor smooth. This part of the seafloor, shown in **Figure 7,** is called the abyssal (uh BIH sul) plain.

Moving Continents Around

The theory of seafloor spreading provides a way to explain how continents move. Continents do not move through the solid mantle or the seafloor. Instead, continents move as the seafloor spreads along a mid-ocean ridge.

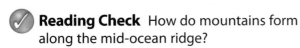

Inquiry MiniLab   20 minutes

How old is the Atlantic Ocean?

If you measure the width of the Atlantic Ocean and you know the rate of seafloor spreading, you can calculate the age of the Atlantic.

1. Use a **ruler** to measure the horizontal distance between a point on the eastern coast of South America and a point on the western coast of Africa on a **world map.** Repeat three times and calculate the average distance in your Science Journal.

2. Use the map's legend to convert the average distance from centimeters to kilometers.

3. If Africa and South America have been moving away from each other at a rate of 2.5 cm per year, calculate the age of the Atlantic Ocean.

Analyze and Conclude

1. **Measure** Did your measurements vary?

2. **Key Concept** How does the age you calculated compare to the breakup of Pangaea 200 million years ago?

Abyssal Plain

Figure 7 The abyssal plain is flat due to an accumulation of sediments far from the ridge.

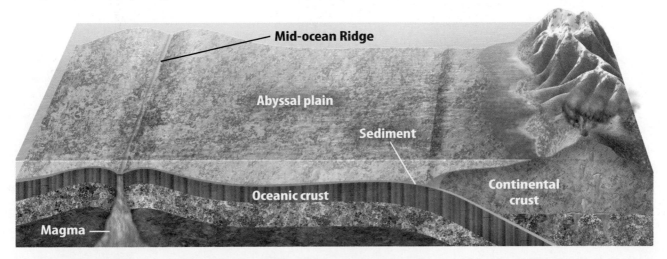

Mid-ocean Ridge

Abyssal plain

Sediment

Oceanic crust

Continental crust

Magma

Visual Check Compare and contrast the topography of a mid-ocean ridge to an abyssal plain.

Development of a Theory

The first evidence used to support seafloor spreading was discovered in rocks on the seafloor. Scientists studied the magnetic signature of minerals in these rocks. To understand this, you need to understand the direction and orientation of Earth's magnetic field and how rocks record magnetic information.

Magnetic Reversals

Recall that the iron-rich, liquid outer core is like a giant magnet that creates Earth's magnetic field. The direction of the magnetic field is not constant. Today's magnetic field, shown in **Figure 8,** is described as having **normal polarity**—*a state in which magnetized objects, such as compass needles, will orient themselves to point north.* Sometimes a **magnetic reversal** *occurs and the magnetic field reverses direction.* The opposite of normal polarity is **reversed polarity**—*a state in which magnetized objects would reverse direction and orient themselves to point south,* as shown in **Figure 8.** Magnetic reversals occur every few hundred thousand to every few million years.

Reading Check Is Earth's magnetic field currently normal or reversed polarity?

Rocks Reveal Magnetic Signature

Basalt on the seafloor contains iron-rich minerals that are magnetic. Each mineral acts like a small magnet. **Figure 9** shows how magnetic minerals align themselves with Earth's magnetic field. When lava erupts from a vent along a mid-ocean ridge, it cools and crystallizes. This permanently records the direction and orientation of Earth's magnetic field at the time of the eruption. Scientists have discovered parallel patterns in the magnetic signature of rocks on either side of a mid-ocean ridge.

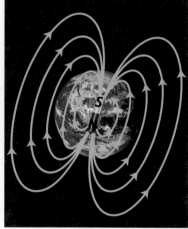

Reversed magnetic field

Normal magnetic field

▲ **Figure 8** Earth's magnetic field is like a large bar magnet. It has reversed direction hundreds of times throughout history.

Figure 9 Iron-rich minerals in cooling lava align with Earth's magnetic field. When Earth's magnetic field changes direction, minerals in fresh lava record a new magnetic signature. ▶

Visual Check Describe the pattern in the magnetic stripes shown in the image to the right.

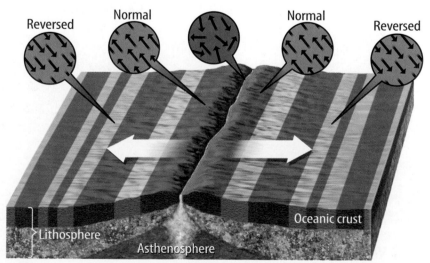

Figure 10 A mirror image in the magnetic stripes on either side of the mid-ocean ridge shows that the crust formed at the ridge is carried away in opposite directions.

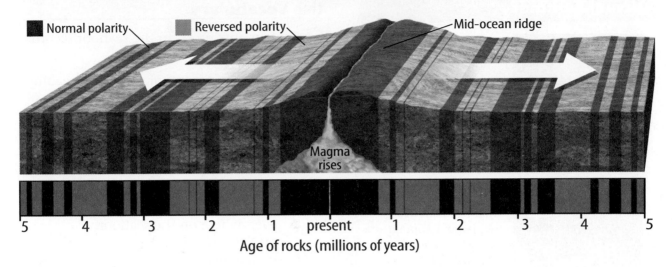

■ Normal polarity   ■ Reversed polarity   Mid-ocean ridge

Magma rises

5   4   3   2   1   present   1   2   3   4   5

Age of rocks (millions of years)

Evidence to Support the Theory

How did scientists prove the theory of seafloor spreading? Scientists studied magnetic minerals in rocks from the seafloor. They used a magnetometer (mag nuh TAH muh tur) to measure and record the magnetic signature of these rocks. These measurements revealed a surprising pattern. Scientists have discovered parallel magnetic stripes on either side of the mid-ocean ridge. Each pair of stripes has a similar composition, age, and magnetic character. Each magnetic stripe in **Figure 10** represents crust that formed and magnetized at a mid-ocean ridge during a period of either **normal** or reversed polarity. The pairs of magnetic stripes confirm that the ocean crust formed at mid-ocean ridges is carried away from the center of the ridges in opposite directions.

ACADEMIC VOCABULARY

normal
(adjective) conforming to a type, standard, or regular pattern

 Reading Check How do magnetic minerals help support the theory of seafloor spreading?

Other measurements made on the seafloor confirm seafloor spreading. By drilling a hole into the seafloor and measuring the temperature beneath the surface, scientists can measure the amount of thermal energy leaving Earth. The measurements show that more thermal energy leaves Earth near mid-ocean ridges than is released from beneath the abyssal plains.

Additionally, sediment collected from the seafloor can be dated. Results show that the sediment closest to the mid-ocean ridge is younger than the sediment farther away from the ridge. Sediment thickness also increases with distance away from the mid-ocean ridge.

FOLDABLES

Make a layered book using two sheets of notebook paper. Use the two pages to record your notes and the inside to illustrate seafloor spreading.

Seafloor Spreading

Lesson 2 Review

Visual Summary

Lava erupts along mid-ocean ridges.

Mid-ocean ridges are large mountain ranges that extend throughout Earth's oceans.

A magnetic reversal occurs when Earth's magnetic field changes direction.

FOLDABLES®

Use your lesson Foldable to review the lesson. Save your Foldable for the project at the end of the chapter.

What do you think NOW?

You first read the statements below at the beginning of the chapter.

3. The seafloor is flat.

4. Volcanic activity occurs only on the seafloor.

Did you change your mind about whether you agree or disagree with the statements? Rewrite any false statements to make them true.

Use Vocabulary

1 Explain how rocks on the seafloor record magnetic reversals over time.

2 Diagram the process of seafloor spreading.

3 Use the term *seafloor spreading* to explain how a mid-ocean ridge forms.

Understand Key Concepts

4 Oceanic crust forms
  A. at mid-ocean ridges.
  B. everywhere on the seafloor.
  C. on the abyssal plains.
  D. by magnetic reversals.

5 Explain why magnetic stripes on the seafloor are parallel to the mid-ocean ridge.

6 Describe how scientists can measure the depth to the seafloor.

Interpret Graphics

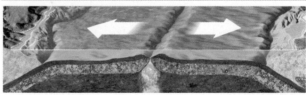

7 Determine Refer to the image above. Where is the youngest crust? Where is the oldest crust?

8 Describe how seafloor spreading helps to explain the continental drift hypothesis.

9 Sequence Information Copy and fill in the graphic organizer below to explain the steps in the formation of a mid-ocean ridge.

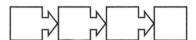

Critical Thinking

10 Infer why magnetic stripes in the Pacific Ocean are wider than in the Atlantic Ocean.

11 Explain why the thickness of seafloor sediments increases with increasing distance from the ocean ridge.

How do rocks on the seafloor vary with age away from a mid-ocean ridge?

Scientists discovered that new ocean crust forms at a mid-ocean ridge and spreads away from the ridge slowly over time. This process is called seafloor spreading. The age of the seafloor is one component that supports this theory.

Materials

vanilla yogurt
berry yogurt

foam board
(10 cm × 4 cm)

waxed paper

plastic spoon

Safety

Do not eat anything used in this lab.

Learn It

Scientists use **models** to represent real-world science. By creating a three-dimensional model of volcanic activity along the Mid-Atlantic Ridge, scientists can model the seafloor spreading process. They can then compare this process to the actual age of the seafloor. In this skill lab, you will investigate how the age of rocks on the seafloor changes with distance away from the ridge.

Million Years B. P.

Try It

1. Read and complete a lab safety form.

2. Lay the sheet of waxed paper flat on the lab table. Place two spoonfuls of vanilla yogurt in a straight line near the center of the waxed paper, leaving it lumpy and full.

3. Lay the two pieces of foam board over the yogurt, leaving a small opening in the middle. Push the foam boards together and down, so the yogurt oozes up and over each of the foam boards.

4. Pull the foam boards apart and add a new row of two spoonfuls of berry yogurt down the middle. Lift the boards and place them partly over the new row. Push them together gently. Observe the outer edges of the new yogurt while you are moving the foam boards together.

5. Repeat step 4 with one more spoonful of vanilla yogurt. Then repeat again with one more spoonful of berry yogurt.

Apply It

6. Compare the map and the model. Where is the Mid-Atlantic Ridge on the map? Where is it represented in your model?

7. Which of your yogurt strips matches today on this map? And millions of years ago?

8. How do scientists determine the ages of different parts of the ocean floor?

9. **Conclude** What happened to the yogurt when you added more?

10. **Key Concept** What happens to the material already on the ocean floor when magma erupts along a mid-ocean ridge?

The Theory of Plate Tectonics

Reading Guide

Key Concepts 🔑
ESSENTIAL QUESTIONS

- What is the theory of plate tectonics?
- What are the three types of plate boundaries?
- Why do tectonic plates move?

Vocabulary

plate tectonics p. 233

lithosphere p. 234

divergent plate boundary p. 235

transform plate boundary p. 235

convergent plate boundary p. 235

subduction p. 235

convection p. 238

ridge push p. 239

slab pull p. 239

g Multilingual eGlossary

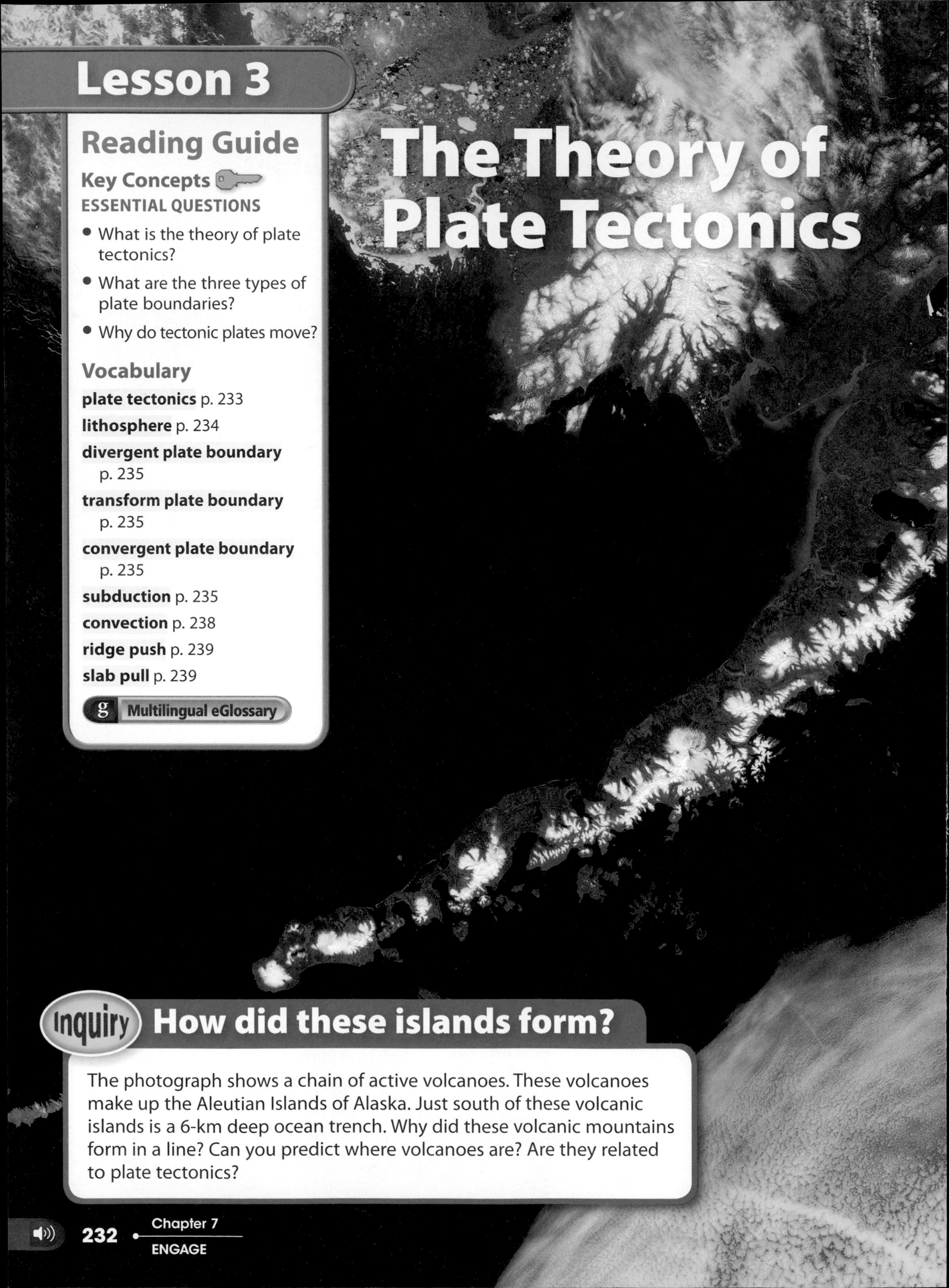

Inquiry How did these islands form?

The photograph shows a chain of active volcanoes. These volcanoes make up the Aleutian Islands of Alaska. Just south of these volcanic islands is a 6-km deep ocean trench. Why did these volcanic mountains form in a line? Can you predict where volcanoes are? Are they related to plate tectonics?

Can you determine density by observing buoyancy?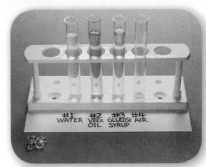

Density is the measure of an object's mass relative to its volume. Buoyancy is the upward force a liquid places on objects that are immersed in it. If you immerse objects with equal densities into liquids that have different densities, the buoyant forces will be different. An object will sink or float depending on the density of the liquid compared to the object. Earth's layers differ in density. These layers float or sink depending on density and buoyant force.

1. Read and complete a lab safety form.

2. Obtain four **test tubes.** Place them in a **test-tube rack.** Add **water** to one test tube until it is ¾ full.

3. Repeat with the other test tubes using **vegetable oil** and **glucose syrup.** One test tube should remain empty.

4. Drop **beads** of equal density into each test tube. Observe what the object does when immersed in each liquid. Record your observations in your Science Journal.

Think About This

1. How did you determine which liquid has the highest density?

2. 🔑 Key Concept What happens when layers of rock with different densities collide?

The Plate Tectonics Theory

When you blow into a balloon, the balloon expands and its surface area also increases. Similarly, if oceanic crust continues to form at mid-ocean ridges and is never destroyed, Earth's surface area should increase. However, this is not the case. The older crust must be destroyed somewhere—but where?

By the late 1960s a more complete theory, called plate tectonics, was proposed. The theory of **plate tectonics** states that *Earth's surface is made of rigid slabs of rock, or plates, that move with respect to each other.* This new theory suggested that Earth's surface is divided into large plates of rigid rock. Each plate moves over Earth's hot and semi-plastic mantle.

 Key Concept Check What is plate tectonics?

Geologists use the word *tectonic* to describe the forces that shape Earth's surface and the rock structures that form as a result. Plate tectonics provides an explanation for the occurrence of earthquakes and volcanic eruptions. When plates separate on the seafloor, earthquakes result and a mid-ocean ridge forms. When plates come together, one plate can dive under the other, causing earthquakes and creating a chain of volcanoes. When plates slide past each other, earthquakes can result.

Earth's Tectonic Plates 🔑

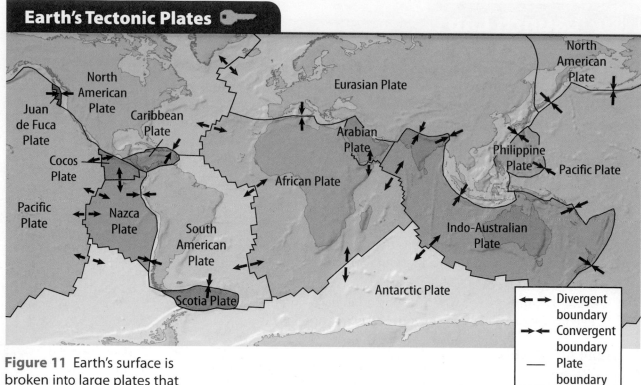

Legend:
- ← → Divergent boundary
- → ← Convergent boundary
- — Plate boundary

Figure 11 Earth's surface is broken into large plates that fit together like pieces of a giant jigsaw puzzle. The arrows show the general direction of movement of each plate.

Tectonic Plates

You read on the previous page that the theory of plate tectonics states that Earth's surface is divided into rigid plates that move relative to one another. These plates are "floating" on top of a hot and semi-plastic mantle. The map in **Figure 11** illustrates Earth's major plates and the boundaries that define them. The Pacific Plate is the largest plate. The Juan de Fuca Plate is one of the smallest plates. It is between the North American and Pacific Plates. Notice the boundaries that run through the oceans. Many of these boundaries mark the positions of the mid-ocean ridges.

Earth's outermost layers are cold and rigid compared to the layers within Earth's interior. *The cold and rigid outermost rock layer is called the* **lithosphere.** It is made up of the crust and the solid, uppermost mantle. The lithosphere is thin below mid-ocean ridges and thick below continents. Earth's tectonic plates are large pieces of lithosphere. These lithospheric plates fit together like the pieces of a giant jigsaw puzzle.

The layer of Earth below the lithosphere is called the asthenosphere (as THEN uh sfihr). This layer is so hot that it behaves like a **plastic** material. This enables Earth's plates to move because the hotter, plastic mantle material beneath them can flow. The interactions between lithosphere and asthenosphere help to explain plate tectonics.

✓ **Reading Check** What are Earth's outermost layers called?

Plate Boundaries

Place two books side by side and imagine each book represents a tectonic plate. A plate boundary exists where the books meet. How many different ways can you move the books with respect to each other? You can pull the books apart, you can push the books together, and you can slide the books past one another. Earth's tectonic plates move in much the same way.

Divergent Plate Boundaries

Mid-ocean ridges are located along divergent plate boundaries. A **divergent plate boundary** *forms where two plates separate.* When the seafloor spreads at a mid-ocean ridge, lava erupts, cools, and forms new oceanic crust. Divergent plate boundaries can also exist in the middle of a continent. They pull continents apart and form rift valleys. The East African Rift is an example of a continental rift.

Transform Plate Boundaries

The famous San Andreas Fault in California is an example of a transform plate boundary. A **transform plate boundary** *forms where two plates slide past each other.* As they move past each other, the plates can get stuck and stop moving. Stress builds up where the plates are "stuck." Eventually, the stress is too great and the rocks break, suddenly moving apart. This results in a rapid release of energy as earthquakes.

Convergent Plate Boundaries

Convergent plate boundaries *form where two plates collide. The denser plate sinks below the more buoyant plate in a process called* **subduction.** The area where a denser plate descends into Earth along a convergent plate boundary is called a subduction zone.

When an oceanic plate and a continental plate collide, the denser oceanic plate subducts under the edge of the continent. This creates a deep ocean trench. A line of volcanoes forms above the subducting plate on the edge of the continent. This process can also occur when two oceanic plates collide. The older and denser oceanic plate will subduct beneath the younger oceanic plate. This creates a deep ocean trench and a line of volcanoes called an island arc.

When two continental plates collide, neither plate is subducted, and mountains such as the Himalayas in southern Asia form from uplifted rock. **Table 1** on the next page summarizes the interactions of Earth's tectonic plates.

 Key Concept Check What are the three types of plate boundaries?

FOLDABLES®

Make a layered book using two sheets of notebook paper. Use it to organize information about the different types of plate boundaries and the features that form there.

Plate Boundaries
Divergent
Convergent
Transform

WORD ORIGIN · · · · · · · · · ·

subduction
from Latin *subductus*, means "to lead under, removal"

Table 1 The direction of motion of Earth's plates creates a variety of features at the boundaries between the plates.

((○)) Concepts in Motion Animation

Table 1 Interactions of Earth's Tectonic Plates ⚷

Plate Boundary	Relative Motion	Example
Divergent plate boundary When two plates separate and create new oceanic crust, a divergent plate boundary forms. This process occurs where the seafloor spreads along a mid-ocean ridge, as shown to the right. This process can also occur in the middle of continents and is referred to as a continental rifting.	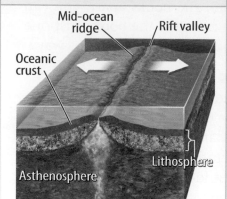	
Transform plate boundary Two plates slide horizontally past one another along a transform plate boundary. Earthquakes are common along this type of plate boundary. The San Andreas Fault, shown to the right, is part of the transform plate boundary that extends along the coast of California.	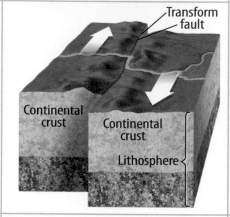	
Convergent plate boundary (ocean-to-continent) When an oceanic and a continental plate collide, they form a convergent plate boundary. The denser plate will subduct. A volcanic mountain, such as Mount Rainier in the Cascade Mountains, forms along the edge of the continent. This process can also occur where two oceanic plates collide, and the denser plate is subducted.	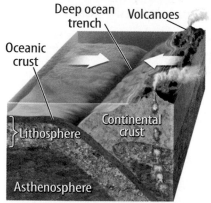	
Convergent plate boundary (continent-to-continent) Convergent plate boundaries can also occur where two continental plates collide. Because both plates are equally dense, neither plate will subduct. Both plates uplift and deform. This creates huge mountains like the Himalayas, shown to the right.	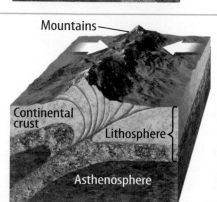	

Evidence for Plate Tectonics

When Wegener proposed the continental drift hypothesis, the technology used to measure how fast the continents move today wasn't yet available. Recall that continents move apart or come together at speeds of a few centimeters per year. This is about the length of a small paperclip.

Today, scientists can measure how fast continents move. A network of satellites orbiting Earth monitors plate motion. By keeping track of the distance between these satellites and Earth, it is possible to locate and determine how fast a tectonic plate moves. This network of satellites is called the Global Positioning System (GPS).

The theory of plate tectonics also provides an explanation for why earthquakes and volcanoes occur in certain places. Because plates are rigid, tectonic activity occurs where plates meet. When plates separate, collide, or slide past each other along a plate boundary, stress builds. A rapid release of energy can result in earthquakes. Volcanoes form where plates separate along a mid-ocean ridge or a continental rift or collide along a subduction zone. Mountains can form where two continents collide. **Figure 12** illustrates the relationship between plate boundaries and the occurrence of earthquakes and volcanoes. Refer back to the lesson opener photo. Find these islands on the map. Are they located near a plate boundary?

 Key Concept Check How are earthquakes and volcanoes related to the theory of plate tectonics?

Figure 12 Notice that most earthquakes and volcanoes occur near plate boundaries.

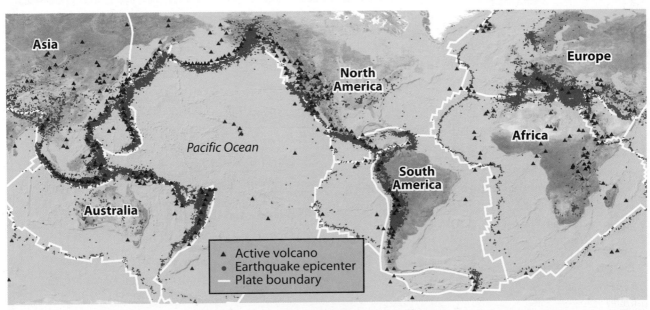

Visual Check Do earthquakes and volcanoes occur anywhere away from plate boundaries?

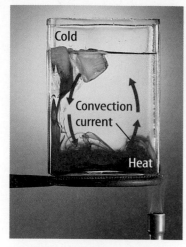

Cold

Convection current

Heat

Figure 13 When water is heated, it expands. Less dense heated water rises because the colder water sinks, forming convection currents.

Review   **Personal Tutor**

Plate Motion

The main objection to Wegener's continental drift hypothesis was that he could not explain why or how continents move. Scientists now understand that continents move because the asthenosphere moves underneath the lithosphere.

Convection Currents

You are probably already familiar with the process of **convection,** *the circulation of material caused by differences in temperature and density.* For example, the upstairs floors of homes and buildings are often warmer. This is because hot air rises while dense, cold air sinks. Look at **Figure 13** to see convection in action.

✓ **Reading Check** What causes convection?

Plate tectonic activity is related to convection in the mantle, as shown in **Figure 14.** Radioactive elements, such as uranium, thorium, and potassium, heat Earth's interior. When materials such as solid rock are heated, they expand and become less dense. Hot mantle material rises upward and comes in contact with Earth's crust. Thermal energy is transferred from hot mantle material to the colder surface above. As the mantle cools, it becomes denser and then sinks, forming a convection current. These currents in the asthenosphere act like a conveyor belt moving the lithosphere above.

☑🔑 **Key Concept Check** Why do tectonic plates move?

Inquiry **MiniLab**

20 minutes

How do changes in density cause motion?

Convection currents drive plate motion. Material near the base of the mantle is heated, which decreases its density. This material then rises to the base of the crust, where it cools, increasing in density and sinking.

❶ Read and complete a lab safety form.

❷ Copy the table to the right into your Science Journal and add a row for each minute. Record your observations.

❸ Pour 100 mL of **carbonated water** or **clear soda** into a **beaker** or a **clear glass.**

❹ Drop five **raisins** into the water. Observe the path that the raisins follow for 5 minutes.

Time Interval	Observations
First minute	
Second minute	
Third minute	

Analyze and Conclude

1. **Observe** Describe each raisin's motion.

2. 🔑 **Key Concept** How does the behavior of the raisin model compare to the motion in Earth's mantle?

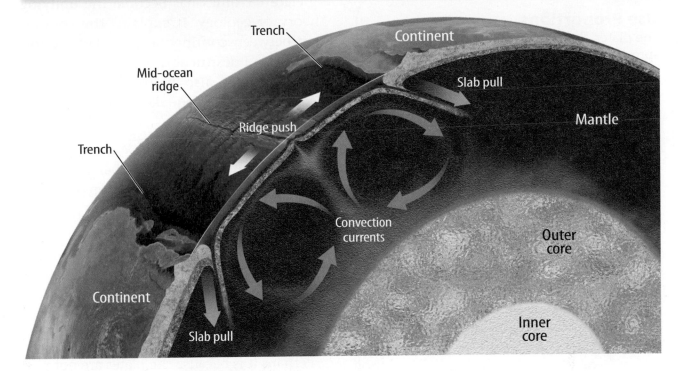

Trench

Continent

Mid-ocean ridge

Slab pull

Ridge push

Mantle

Trench

Convection currents

Outer core

Continent

Inner core

Slab pull

Forces Causing Plate Motion

How can something as massive as the Pacific Plate move? **Figure 14** shows the three forces that interact to cause plate motion. Scientists still debate over which of these forces has the greatest effect on plate motion.

Basal Drag Convection currents in the mantle produce a force that causes motion called basal drag. Notice in **Figure 14** how convection currents in the asthenosphere circulate and drag the lithosphere similar to the way a conveyor belt moves items along at a supermarket checkout.

Ridge Push Recall that mid-ocean ridges have greater elevation than the surrounding seafloor. Because mid-ocean ridges are higher, gravity pulls the surrounding rocks down and away from the ridge. *Rising mantle material at mid-ocean ridges creates the potential for plates to move away from the ridge with a force called* **ridge push**. Ridge push moves lithosphere in opposite directions away from the mid-ocean ridge.

Slab Pull As you read earlier in this lesson, when tectonic plates collide, the denser plate will sink into the mantle along a subduction zone. This plate is called a slab. Because the slab is old and cold, it is denser than the surrounding mantle and will sink. *As a slab sinks, it pulls on the rest of the plate with a force called* **slab pull**. Scientists are still uncertain about which force has the greatest influence on plate motion.

Figure 14 Convection occurs in the mantle underneath Earth's tectonic plates. Three forces act on plates to make them move: basal drag from convection currents, ridge push at mid-ocean ridges, and slab pull from subducting plates.

✓ **Visual Check** What is happening to a plate that is undergoing slab pull?

Use Proportions

The plates along the Mid-Atlantic Ridge spread at an average rate of 2.5 cm/y. How long will it take the plates to spread 1 m? Use proportions to find the answer.

1 **Convert the distance to the same unit.**

$$1 \text{ m} = 100 \text{ cm}$$

2 **Set up a proportion:**

$$\frac{2.5 \text{ cm}}{1 \text{ y}} = \frac{100 \text{ cm}}{x \text{ y}}$$

3 **Cross multiply and solve for *x* as follows:**

$$2.5 \text{ cm} \times x\text{y} = 100 \text{ cm} \times 1 \text{ y}$$

4 **Divide both sides by 2.5 cm.**

$$x = \frac{100 \text{ cm y}}{2.5 \text{ cm}}$$

$$x = 40 \text{ y}$$

Practice

The Eurasian plate travels the slowest, at about 0.7 cm/y. How long would it take the plate to travel 3 m?

$$(1 \text{ m} = 100 \text{ cm})$$

 Review

- **Math Practice**
- **Personal Tutor**

Vertical mantle section

Slab

Velocity of seismic waves

Slow ▬▬▬▬▬▬▬▬▬ Fast

A Theory in Progress

Plate tectonics has become the unifying theory of geology. It explains the connection between continental drift and the formation and destruction of crust along plate boundaries. It also helps to explain the occurrence of earthquakes, volcanoes, and mountains.

The investigation that Wegener began nearly a century ago is still being revised. Several unanswered questions remain.

- Why is Earth the only planet in the solar system that has plate tectonic activity? Different hypotheses have been proposed to explain this. Extrasolar planets outside our solar system are also being studied.

- Why do some earthquakes and volcanoes occur far away from plate boundaries? Perhaps it is because the plates are not perfectly rigid. Different thicknesses and weaknesses exist within the plates. Also, the mantle is much more active than scientists originally understood.

- What forces dominate plate motion? Currently accepted models suggest that convection currents occur in the mantle. However, there is no way to measure or observe them.

- What will scientists investigate next? **Figure 15** shows an image produced by a new technique called anisotropy that creates a 3-D image of seismic wave velocities in a subduction zone. This developing technology might help scientists better understand the processes that occur within the mantle and along plate boundaries.

Reading Check Why does the theory of plate tectonics continue to change?

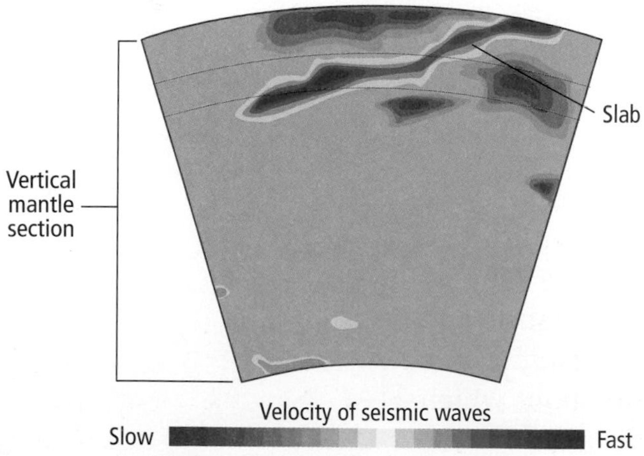

Figure 15 Seismic waves were used to produce this tomography scan. These colors show a subducting plate. The blue colors represent rigid materials with faster seismic wave velocities.

Lesson 3 Review

Visual Summary

Lithosphere

Tectonic plates are made of cold and rigid slabs of rock.

Mantle convection—the circulation of mantle material due to density differences—drives plate motion.

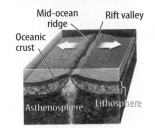

Mid-ocean ridge   Rift valley
Oceanic crust
Asthenosphere   Lithosphere

The three types of plate boundaries are divergent, convergent, and transform boundaries.

FOLDABLES

Use your lesson Foldable to review the lesson. Save your Foldable for the project at the end of the chapter.

What do you think NOW?

You first read the statements below at the beginning of the chapter.

5. Continents drift across a molten mantle.

6. Mountain ranges can form when continents collide.

Did you change your mind about whether you agree or disagree with the statements? Rewrite any false statements to make them true.

Use Vocabulary

1. The theory that proposes that Earth's surface is broken into moving, rigid plates is called _____.

Understand Key Concepts

2. **Compare and contrast** the geological activity that occurs along the three types of plate boundaries.

3. **Explain** why mantle convection occurs.

4. Tectonic plates move because of
   A. convection currents.
   B. Earth's increasing size.
   C. magnetic reversals.
   D. volcanic activity.

Interpret Graphics

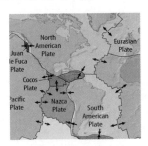

North American Plate   Eurasian Plate
Juan de Fuca Plate
Cocos Plate
Pacific Plate   Nazca Plate   South American Plate

5. **Identify** Name the type of boundary between the Eurasian Plate and the North American Plate and between the Nazca Plate and South American Plate.

6. **Determine Cause and Effect** Copy and fill in the graphic organizer below to list the cause and effects of convection currents.

Critical Thinking

7. **Explain** why earthquakes occur at greater depths along convergent plate boundaries.

Math Skills ×÷   Review — Math Practice

8. Two plates in the South Pacific separate at an average rate of 15 cm/y. How far will they have separated after 5,000 years?

Movement of Plate Boundaries

Earth's surface is broken into 12 major tectonic plates. Wherever these plates touch, one of four events occurs. The plates may collide and crumple or fold to make mountains. One plate may subduct under another, forming volcanoes. They may move apart and form a mid-ocean ridge, or they may slide past each other causing an earthquake. This investigation models plate movements.

Materials

graham crackers

waxed paper (four 10×10-cm squares)

dropper

frosting

plastic spoon

Safety

Question

What happens where two plates come together?

Procedure

Part I

1. Read and complete a lab safety form.
2. Obtain the materials from your teacher.
3. Break a graham cracker along the perforation line into two pieces.
4. Lay the pieces side by side on a piece of waxed paper.
5. Slide crackers in opposite directions so that the edges of the crackers rub together.

Part II

6. Place two new graham crackers side by side but not touching.
7. In the space between the crackers, add several drops of water.
8. Slide the crackers toward each other and observe what happens.

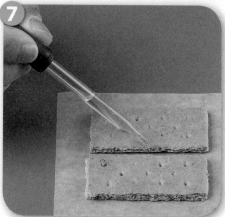

Part III

9. Place a spoonful of frosting on the waxed-paper square.
10. Place two graham crackers on top of the frosting so that they touch.
11. Push the crackers down and spread them apart in one motion.

Analyze and Conclude

12 Analyze the movement of the crackers in each of your models.

Part I

13 What type of plate boundary do the graham crackers in this model represent?

14 What do the crumbs in the model represent?

15 Did you feel or hear anything when the crackers moved past each other? Explain.

16 How does this model simulate an earthquake?

Part II

17 What does the water in this model represent?

18 What type of plate boundary do the graham crackers in this model represent?

19 Why didn't one graham cracker slide beneath the other in this model?

Part III

20 What type of plate boundary do the graham crackers in this model represent?

21 What does the frosting represent?

22 What shape does the frosting create when the crackers move?

23 What is the formation formed from the crackers and frosting?

Communicate Your Results

Create a flip book of one of the boundaries to show a classmate who was absent. Show how each boundary plate moves and the results of those movements.

 **Extension**

Place a graham cracker and a piece of cardboard side by side. Slide the two pieces toward each other. What type of plate boundary does this model represent? How is this model different from the three that you observed in the lab?

Lab Tips

☑ Use fresh graham crackers.

☑ Slightly heat frosting to make it more fluid for experiments.

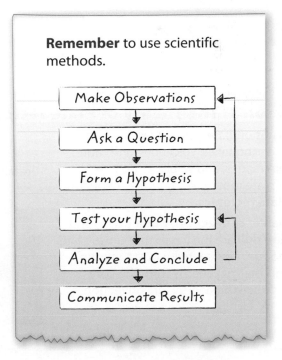

Remember to use scientific methods.

Make Observations → Ask a Question → Form a Hypothesis → Test your Hypothesis → Analyze and Conclude → Communicate Results

Chapter 7 Study Guide

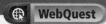

 THE BIG IDEA The theory of plate tectonics states that Earth's lithosphere is broken up into rigid plates that move over Earth's surface.

Key Concepts Summary ⚷

	Vocabulary

Lesson 1: The Continental Drift Hypothesis

- The puzzle piece fit of continents, fossil evidence, climate, rocks, and mountain ranges supports the hypothesis of **continental drift.**
- Scientists were skeptical of continental drift because Wegener could not explain the mechanism for movement.

Pangaea p. 217
continental drift p. 217

Lesson 2: Development of a Theory

- **Seafloor spreading** provides a mechanism for continental drift.
- Seafloor spreading occurs at **mid-ocean ridges.**
- Evidence of **magnetic reversal** in rock, thermal energy trends, and the discovery of seafloor spreading all contributed to the development of the theory of plate tectonics.

mid-ocean ridge p. 225
seafloor spreading p. 226
normal polarity p. 228
magnetic reversal p. 228
reversed polarity p. 228

Lesson 3: The Theory of Plate Tectonics

- Types of plate boundaries, the location of earthquakes, volcanoes, and mountain ranges, and satellite measurement of plate motion support the theory of **plate tectonics.**
- Mantle **convection, ridge push,** and **slab pull** are the forces that cause plate motion. Radioactivity in the mantle and thermal energy from the core produce the energy for convection.

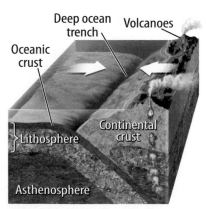

Deep ocean trench — Volcanoes
Oceanic crust
Lithosphere
Continental crust
Asthenosphere

plate tectonics p. 233
lithosphere p. 234
divergent plate boundary p. 235
transform plate boundary p. 235
convergent plate boundary p. 235
subduction p. 235
convection p. 238
ridge push p. 239
slab pull p. 239

FOLDABLES® Chapter Project

Assemble your lesson Foldables as shown to make a Chapter Project. Use the project to review what you have learned in this chapter.

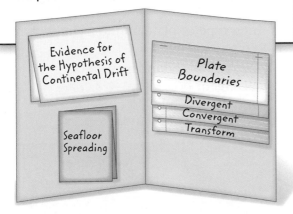

Evidence for the Hypothesis of Continental Drift

Plate Boundaries
Divergent
Convergent
Transform

Seafloor Spreading

Use Vocabulary

1 The process in which hot mantle rises and cold mantle sinks is called _____.

2 What is the plate tectonics theory?

3 What was Pangaea?

4 Identify the three types of plate boundaries and the relative motion associated with each type.

5 Magnetic reversals occur when _____.

6 Explain seafloor spreading in your own words.

Link Vocabulary and Key Concepts

Concepts in Motion Interactive Concept Map

Copy this concept map, and then use vocabulary terms from the previous page to complete the concept map.

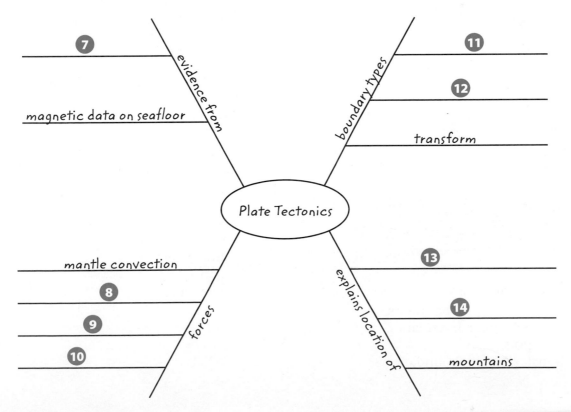

7

magnetic data on seafloor
— evidence from

11

12

transform
— boundary types

Plate Tectonics

mantle convection
8
9
10
— forces

13

14

mountains
— explains location of

Understand Key Concepts 🔑

1 Alfred Wegener proposed the _____ hypothesis.
   A. continental drift
   B. plate tectonics
   C. ridge push
   D. seafloor spreading

2 Ocean crust is
   A. made from submerged continents.
   B. magnetically produced crust.
   C. produced at the mid-ocean ridge.
   D. produced at all plate boundaries.

3 What technologies did scientists NOT use to develop the theory of seafloor spreading?
   A. echo-sounding measurements
   B. GPS (global positioning system)
   C. magnetometer measurements
   D. seafloor thickness measurements

4 The picture below shows Pangaea's position on Earth approximately 280 million years ago. Where did geologists discover glacial features associated with a cooler climate?
   A. Antarctica
   B. Asia
   C. North America
   D. South America

Pangaea

5 Mid-ocean ridges are associated with
   A. convergent plate boundaries.
   B. divergent plate boundaries.
   C. hot spots.
   D. transform plate boundaries.

6 Two plates of equal density form mountain ranges along
   A. continent-to-continent convergent boundaries.
   B. ocean-to-continent convergent boundaries.
   C. divergent boundaries.
   D. transform boundaries.

7 Which type of plate boundary is shown in the figure below?
   A. convergent boundary
   B. divergent boundary
   C. subduction zone
   D. transform boundary

8 What happens to Earth's magnetic field over time?
   A. It changes polarity.
   B. It continually strengthens.
   C. It stays the same.
   D. It weakens and eventually disappears.

9 Which of Earth's outermost layers includes the crust and the upper mantle?
   A. asthenosphere
   B. lithosphere
   C. mantle
   D. outer core

Critical Thinking

10 **Evaluate** The oldest seafloor in the Atlantic Ocean is located closest to the edge of continents, as shown in the image below. Explain how this age can be used to figure out when North America first began to separate from Europe.

11 **Examine** the evidence used to develop the theory of plate tectonics. How has new technology strengthened the theory?

12 **Explain** Sediments deposited by glaciers in Africa are surprising because Africa is now warm. How does the hypothesis of continental drift explain these deposits?

13 **Draw** a diagram to show subduction of an oceanic plate beneath a continental plate along a convergent plate boundary. Explain why volcanoes form along this type of plate boundary.

14 **Infer** Warm peanut butter is easier to spread than cold peanut butter. How does knowing this help you understand why the mantle is able to deform in a plastic manner?

Writing in Science

15 **Predict** If continents continue to move in the same direction over the next 200 million years, how might the appearance of landmasses change? Write a paragraph to explain the possible positions of landmasses in the future. Based on your understanding of the plate tectonic theory, is it possible that new supercontinents will form in the future?

REVIEW THE BIG IDEA

16 What is the theory of plate tectonics? Distinguish between continental drift, seafloor spreading, and plate tectonics. What evidence was used to support the theory of plate tectonics?

17 Use the image below to interpret how the theory of plate tectonics helps to explain the formation of huge mountains like the Himalayas.

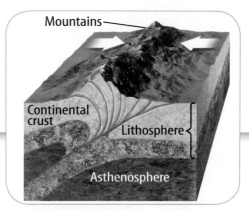

Mountains

Continental crust

Lithosphere

Asthenosphere

Math Skills ×÷

Review — Math Practice

Use Proportions

18 Mountains on a convergent plate boundary may grow at a rate of 3 mm/y. How long would it take a mountain to grow to a height of 3,000 m? (1 m = 1,000 mm)

19 The North American Plate and the Pacific Plate have been sliding horizontally past each other along the San Andreas fault zone for about 10 million years. The plates move at an average rate of about 5 cm/y.

   a. How far have the plates traveled, assuming a constant rate, during this time?

   b. How far has the plate traveled in kilometers? (1 km = 100,000 cm)

Record your answers on the answer sheet provided by your teacher or on a sheet of paper.

Multiple Choice

Use the diagram below to answer questions 1 and 2.

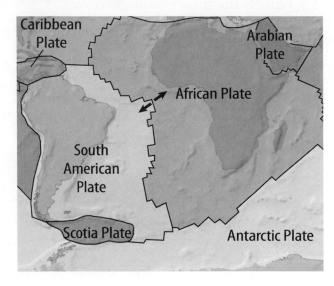

1 In the diagram above, what does the irregular line between tectonic plates represent?

  A abyssal plain

  B island chain

  C mid-ocean ridge

  D polar axis

2 What do the arrows indicate?

  A magnetic polarity

  B ocean flow

  C plate movement

  D volcanic eruption

3 What evidence helped to support the theory of seafloor spreading?

  A magnetic equality

  B magnetic interference

  C magnetic north

  D magnetic polarity

4 Which plate tectonic process creates a deep ocean trench?

  A conduction

  B deduction

  C induction

  D subduction

5 What causes plate motion?

  A convection in Earth's mantle

  B currents in Earth's oceans

  C reversal of Earth's polarity

  D rotation on Earth's axis

6 New oceanic crust forms and old oceanic crust moves away from a mid-ocean ridge during

  A continental drift.

  B magnetic reversal.

  C normal polarity.

  D seafloor spreading.

Use the diagram below to answer question 7.

7 What is the name of Alfred Wegener's ancient supercontinent pictured in the diagram above?

  A Caledonia

  B continental drift

  C *Glossopteris*

  D Pangaea

Use the diagram below to answer question 8.

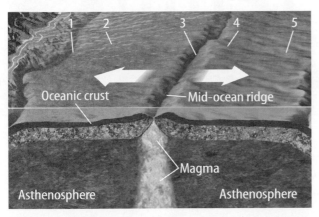

8 The numbers in the diagram represent sea-floor rock. Which represent the oldest rock?

A 1 and 5

B 2 and 4

C 3 and 4

D 4 and 5

9 Which part of the seafloor contains the thickest sediment layer?

A abyssal plain

B deposition band

C mid-ocean ridge

D tectonic zone

10 What type of rock forms when lava cools and crystallizes on the seafloor?

A a fossil

B a glacier

C basalt

D magma

Constructed Response

Use the table below to answer questions 11 and 12.

Plate Boundary	Location

11 In the table above, identify the three types of plate boundaries. Then describe a real-world location for each type.

12 Create a diagram to show plate motion along one type of plate boundary. Label the diagram and draw arrows to indicate the direction of plate motion.

13 Identify and explain all the evidence that Wegener used to help support his continental drift hypothesis.

14 Why was continental drift so controversial during Alfred Wegener's time? What explanation was necessary to support his hypothesis?

15 How did scientists prove the theory of seafloor spreading?

16 If new oceanic crust constantly forms along mid-ocean ridges, why isn't Earth's total surface area increasing?

NEED EXTRA HELP?																
If You Missed Question...	1	2	3	4	5	6	7	8	9	10	11	12	13	14	15	16
Go to Lesson...	3	3	2	3	3	2	1	2	2	2	3	3	1	1	2	3

Chapter 8

Earth Dynamics

THE BIG IDEA How is Earth's surface shaped by plate motion?

Inquiry **Why is Mount Everest different?**

You might think that seashells are only found near oceans. But some of the rocks in Mount Everest contain seashells from the ocean floor!

• How do you think seashells got to the top of Mount Everest?

• What are the different ways tectonic plates move to make mountains such as these—or deep-sea trenches, valleys, and plateaus?

• How is Earth's surface shaped by plate motion?

Get Ready to Read

What do you think?

Before you read, decide if you agree or disagree with each of these statements. As you read this chapter, see if you change your mind about any of the statements.

1 Forces created by plate motion are small and do not deform or break rocks.

2 Plate motion causes only horizontal motion of continents.

3 New landforms are created only at plate boundaries.

4 The tallest and deepest landforms are created at plate boundaries.

5 Metamorphic rocks formed deep below Earth's surface sometimes can be located near the tops of mountains.

6 Mountain ranges can form over long periods of time through repeated collisions between plates.

7 The centers of continents are flat and old.

8 Continents are continually shrinking because of erosion.

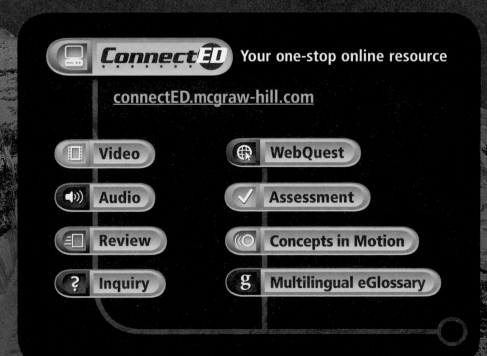

ConnectED Your one-stop online resource

connectED.mcgraw-hill.com

- Video
- Audio
- Review
- Inquiry
- WebQuest
- Assessment
- Concepts in Motion
- Multilingual eGlossary

Lesson 1

Forces That Shape Earth

Reading Guide

Key Concepts
ESSENTIAL QUESTIONS

- How do continents move?
- What forces can change rocks?
- How does plate motion affect the rock cycle?

Vocabulary

isostasy p. 254

subsidence p. 255

uplift p. 255

compression p. 255

tension p. 255

shear p. 255

strain p. 256

g Multilingual eGlossary

Video Science Video

Inquiry Can rocks talk?

This campsite in Thingvellir, Iceland, can tell a story about Earth if you ask the right questions. Why is this cliff next to a flat, grassy valley? How did it get like this? Has it always been this way? You can find some answers by looking at the forces that shape Earth.

Do rocks bend?

As Earth's continents move, rocks get smashed between them and bend or break. Land can take on different shapes, depending on the temperature and composition of the rocks and the size and direction of the force.

1. Read and complete a lab safety form.
2. Spread out a **paper towel** on your work area, and place an unwrapped **candy bar** on the paper towel.
3. Gently pull on the edges of your candy bar. Observe any changes to the candy bar. Draw your observations in your Science Journal.
4. Reassemble your candy bar and gently squeeze the two ends of your candy bar together. Draw your observations.

Think About This

1. How are the results of pulling and pushing different?

2. What would be different if the candy bar were warm? What if it were cold?

3. 🔑 **Key Concept** What kinds of forces do you think can change rocks?

Plate Motion

How far is your school from the nearest large mountain? If you live in the west or along the east coast of the United States, you are probably close to mountains. In contrast, the central region of the United States is flat. Why are these regions so different?

The Rocky Mountains in the west are high and have sharp peaks, but the Appalachian Mountains in the east are lower and gently rounded, as shown in **Figure 1.**

 Reading Check How are the Rocky Mountains different from the Appalachian Mountains?

Mountains do not last forever. Weathering and erosion gradually wear them down. The Appalachian Mountains are shorter and smoother than the Rocky Mountains because they are older. They formed hundreds of millions of years ago. The Rockies formed just 50 to 100 million years ago.

Mountain ranges are produced by plate tectonics. The theory of plate tectonics states that Earth's surface is broken into rigid plates that move horizontally on Earth's more fluid upper mantle. Mountains and valleys form where plates collide, move away from each other, or slide past each other.

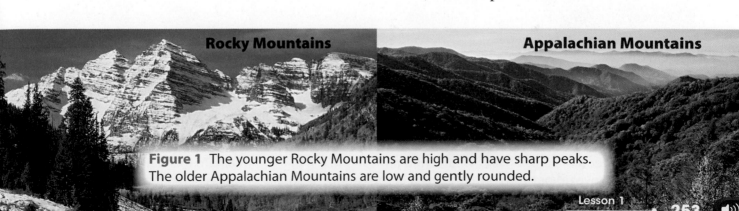

Rocky Mountains

Appalachian Mountains

Figure 1 The younger Rocky Mountains are high and have sharp peaks. The older Appalachian Mountains are low and gently rounded.

Vertical Motion

To understand how massive pieces of Earth can rise vertically and form mountainous regions, you need to understand the forces that produce vertical motion.

Balance in the Mantle

Think of an iceberg floating in water. The iceberg floats with its top above the water, but most of it is under the surface of the water, as shown in **Figure 2.** It floats this way because ice is less dense than water and because the mass of the ice equals the mass of the water it displaces, or pushes out of the way.

Similarly, continents rise above the seafloor because continental crust is made of rocks that are less dense than Earth's mantle. Continental crust displaces some of the mantle below it until an equilibrium, or balance, is reached. **Isostasy** (i SAHS tuh see) *is the equilibrium between continental crust and the denser mantle below it.* A continent floats on top of the mantle because the mass of the continent is equal to the mass of the mantle it displaces. Mountains act the same way on a smaller scale.

 Reading Check What is isostasy?

Continental crust changes over time due to plate tectonics and erosion. If a part of the continental crust becomes thicker, it sinks deeper into the mantle, as shown in **Figure 3.** But it also rises higher until a balance is reached. This is why mountains are taller than the continental crust around them. Although the mountain is massive, it is still less dense than the mantle, so it "floats." Below Earth's surface, the mountain extends deep into the mantle. Above Earth's surface, the mountain rises above the surrounding continental crust. As illustrated in **Figure 3,** as a mountain erodes, the continental crust rises.

Figure 2 The massive lower portion of an iceberg is under water. Similarly, the root of a mountain extends deep into the mantle.

WORD ORIGIN

isostasy
from Greek *iso*, means "equal"; and Greek *stasy*, means "standing"

Maintaining Balance  Review   Personal Tutor

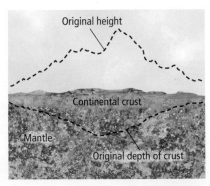

Figure 3 Over time, erosion and weathering remove the top of a mountain. To maintain isostasy, continents move up or down until the mass of the continent equals the mass of mantle it displaces.

Subsidence and Uplift

Much of North America was covered by glaciers more than 1 km thick 20,000 years ago. The weight of the ice pushed the crust downward into the mantle, as shown in **Figure 4.** *The downward vertical motion of Earth's surface is called* **subsidence.** When the ice melted and the water ran off, the isostatic balance was upset again. In response, the crust moved upward. *The upward vertical motion of Earth's surface is called* **uplift.** In the center of Hudson Bay in Canada, the land surface is still rising 1 cm each year as it moves toward isostatic balance.

 Key Concept Check What can cause Earth's surface to move up or down?

Horizontal Motion

Find a small rock and squeeze it. You've just applied force to the rock. Did its shape change? Did it break? Horizontal motion at plate boundaries applies much greater forces to rocks. Forces at plate boundaries are strong enough to break rocks or change the shape of rocks. The same forces also can form mountains.

Types of Stress

Stress is the force acting on a surface. There are three types of stress, as illustrated in **Figure 5.** *Squeezing stress is* **compression.** *Stress that pulls something apart is* **tension.** *Parallel forces acting in opposite directions are* **shear.** These are all stresses that can change rock as plates move horizontally.

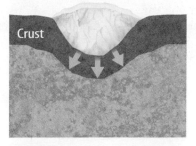

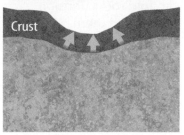

Figure 4 The weight of a glacier pushes down on the land. When the glacier melts, the land rises until isostasy is restored.

Use a sheet of paper to make a three-tab book. Label the tabs as illustrated, and describe the forces that shape Earth.

Stresses

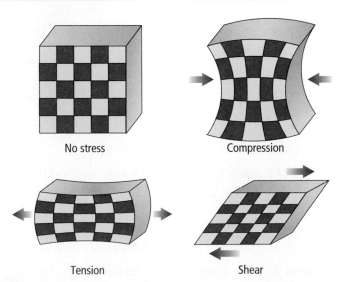

Figure 5 Compression, tension, and shear stress cause rocks to change shape.

Figure 6 Compression can fold rocks. Tension can stretch them. Whether rocks go back to their original shape depends on the type of strain.

 Visual Check Which of these two illustrations shows tension?

SCIENCE USE V. COMMON USE

plastic
Science Use capable of being molded

Common Use a commonly used synthetic material

Types of Strain

Rocks can change when stress acts on them. *A change in the shape of rock caused by stress is called* **strain.** There are two main types of strain.

Elastic strain does not permanently change, or deform, rocks. When stress is removed, rocks return to their original shapes. Elastic strain occurs when stresses are small or rocks are very strong. **Plastic** strain creates a permanent change in shape. Even if the stress is removed, the rocks do not go back to their original shapes. Plastic strain occurs when rocks are weak or hot.

✓ **Reading Check** Which type of strain permanently changes rocks?

Deformation in the Crust

In the hotter lower crust and upper mantle, rocks tend to deform plastically like putty. As illustrated in **Figure 6,** compression thickens and folds layers of rock. Tension stretches and thins layers of rock. In the colder, upper part of the crust, rocks can break before they deform plastically. When strain breaks rocks rather than just changing their shape, it is called failure. When rocks fail, fractures—or faults—form.

🔑 **Key Concept Check** What can cause rocks to thicken or fold?

Inquiry MiniLab

10 minutes

What will happen?

If enough force is put on a rock, it will begin to strain, or change shape. Depending on the nature of the force and the rock, sometimes the rock will bend and sometimes it will break.

1. Read and complete a lab safety form.

2. Knead a piece of **putty,** and pull it apart slowly. Shape the putty into an oval ball. Try to pull it apart quickly. Record your observations in your Science Journal.

3. Shape your putty into an oval. Put your putty in a warm water bath for 2 min. Pull it apart. Record your observations.

4. Shape the putty into an oval shape. Put the putty in an **ice water** bath for 2 min. Pull it apart. Record your observations.

5. Try to break your putty by pulling on it and by pushing on it. Record your observations in your Science Journal.

Analyze and Conclude

1. **Summarize** the effects of rate of strain, temperature, and type of stress on the putty.

2. 🔑 **Key Concept** Relate your experience with your putty model to the forces that can change rocks and to the conditions in Earth that will cause them to change.

Plate Tectonics and the Rock Cycle

Although it might seem as if rocks are always the same, rocks are moving around—usually very slowly. Rocks never stop moving through the rock cycle, as illustrated in **Figure 7.** The theory of plate tectonics combined with uplift and subsidence explain why there is a rock cycle on Earth.

The forces that cause plate tectonics produce horizontal motion. Isostasy results in vertical motion within continents. Together, plate motion, uplift, and subsidence keep rocks moving through the rock cycle.

Uplift brings metamorphic and igneous rocks from deep in the crust up to the surface. At the surface, erosion breaks down rocks into sediment. Sediment gets buried by still more sediment. Buried sediment becomes sedimentary rocks. Pressure and temperature increase as rocks are buried, and eventually sedimentary rocks become metamorphic rocks. Subduction takes all types of rocks deep into Earth, where they can melt and create new igneous or metamorphic rocks.

Key Concept Check How does plate motion affect the rock cycle?

The Rock Cycle

 **Concepts in Motion** Animation

Figure 7 Horizontal tectonic motion and vertical motion by uplift and subsidence help move rocks through the rock cycle.

Visual Check What happens to eroded sediment?

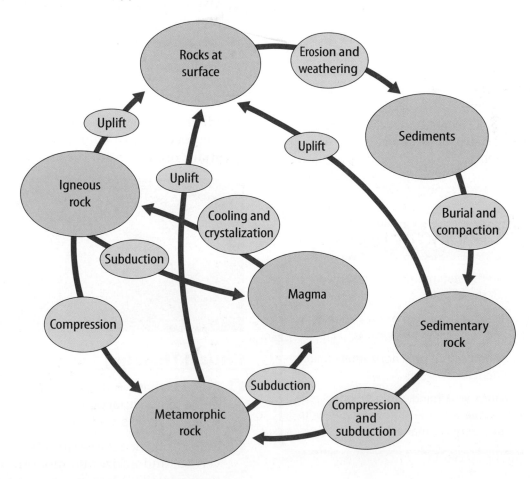

Visual Summary

As a mountain is eroded away, the continent will rise until isostatic balance is restored.

Different types of stress change rocks in different ways.

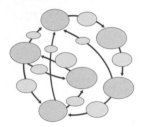

Horizontal and vertical motions are part of what keep rocks moving through the rock cycle.

FOLDABLES

Use your lesson Foldable to review the lesson. Save your Foldable for the project at the end of the chapter.

What do you think NOW?

You first read the statements below at the beginning of the chapter.

1. Forces created by plate motion are small and do not deform or break rocks.

2. Plate motion causes only horizontal motion of continents.

Did you change your mind about whether you agree or disagree with the statements? Rewrite any false statements to make them true.

Use Vocabulary

1 The opposite of uplift is _____.

2 The balance between the crust and the mantle below it is _____.

3 **Explain** the difference between the two types of strain.

Understand Key Concepts 🗝

4 **Describe** what happens to the elevation of the land surface when crust thickens.

5 **Name** one result of rock failure.

6 Which type of deformation is produced by compression of plastic crust?
   A. failure   C. folds
   B. faults   D. shear

Interpret Graphics

7 **List** Copy the graphic organizer below, and use it to show three types of stress.

Stress

8 **Identify** the type of stress that deformed the rocks shown below. What kind of strain resulted from the stress?

Critical Thinking

9 **Predict** what would happen to the height of the land surface of Antarctica if the ice sheet started to melt.

10 **Reflect** on the relationship between vertical and horizontal motion. How are they related? When are they not related?

Materials

assorted
weights

cardboard

ruler

stopwatch

putty

Safety

Can you measure how stress deforms putty?

Scientists who study rocks study how the stress, or force applied to a rock, causes the rock to change shape, or deform. Because rocks are hard and they only deform with enormous forces at slow speeds, special equipment is needed to study rocks. In this lab, you will apply forces to putty and take measurements similar to ones scientists take on rocks.

Learn It

A scientist makes many decisions before beginning an investigation. Some decisions involve figuring out how to make the needed measurements. In this lab, you will **design** instruments and an **experiment** to measure stress and deformation of putty.

Try It

1. Read and complete a lab safety form.

2. Determine how you will use the materials provided to measure stress, deformation, and the time it takes to deform the putty.

3. Write a procedure in your Science Journal, and ask your teacher to approve your plan.

4. Test your procedures. Modify them if necessary.

5. Collect your data, and record them in a data table such as the one shown below.

Apply It

6. **Summarize** the relationship between stress and rate of deformation.

7. 🔑 **Key Concept** Relate the deformation of putty to how forces change rocks.

	Stress Applied	Measured Deformation	Measured Time	Rate of Deformation (deformation / time)
Trial 1				
Trial 2				

Landforms at Plate Boundaries

Reading Guide

Key Concepts 🔑
ESSENTIAL QUESTIONS

- What features form where two plates converge?
- What features form where two plates diverge?
- What features form where two plates slide past each other?

Vocabulary

ocean trench p. 262
volcanic arc p. 263
transform fault p. 265
fault zone p. 265

g **Multilingual eGlossary**

Inquiry What happened here?

What tore this landscape apart? Have you ever seen a place like this? Probably not, because places like this are usually under the ocean! Whether it is under the ocean or on dry land, there's a lot of action at plate boundaries.

What happens when tectonic plates collide?

As Earth's continents move, tectonic plates can come together, pull apart, or slide past each other. Each of these interactions produces different landforms.

1. Using **construction paper** and **index cards,** set up a model plate boundary. Draw your model in your Science Journal.

2. Label the two tectonic plates, the fault between them, and the mantle. Title it *Before Stress.*

3. Model tension by pulling the plates apart. Draw and label your model. Title it *Tension.*

4. Model shear by sliding one plate forward and the other backward. Draw and label your model. Title it *Shear.*

5. Model compression by pushing the plates together. Experiment until you get two different results. Draw and label both results. Title it *Compression.*

Think About This

1. What might happen if compression and shear occurred together?

2. **Key Concept** How are the features that form under the different types of stresses different? How are they similar?

Landforms Created by Plate Motion

Tectonic plates move slowly, only 1–9 cm per year. But these massive, slow-moving plates have so much force they can build tall mountains, form deep valleys, and rip Earth's surface apart.

Compression, tension, and shear stresses are at work at plate boundaries. Each type of stress produces different types of landforms. For example, the San Andreas Fault on the west coast of the United States is the result of shear stresses where plates move past each other. Tall mountains, such as the Ural Mountains shown in **Figure 8,** are created by compression stresses where plates collide.

Reading Check How fast do tectonic plates move?

Figure 8 The Ural Mountains are the result of a collision that started about 250 million years ago between landmasses that are now the continents of Europe and Asia.

Figure 9 Three stages in the growth of the Himalayas are illustrated. The plates beneath India and Asia started colliding almost 50 million years ago and continue colliding today. Because the plates are still colliding, the Himalayas grow a few millimeters each year due to compression.

🔍 **Visual Check** Which two landforms collided?

Himalayas
Eurasian Plate

Eurasian Plate
Indian Plate
Oceanic sediments

Eurasian Plate
Indian Plate
Oceanic sediments

Indian Plate
Oceanic sediments

FOLDABLES®

Use a sheet of paper to make a vertical three-tab book. As you read, describe how different features form at plate boundaries. Include specific examples of each landform.

Rifts and Ridges

Volcanic Arc

Fracture Zones

Landforms Created by Compression

The largest landforms on Earth are produced by compression at convergent plate boundaries. The types of landforms that form depend on whether the plates are oceanic or continental.

Mountain Ranges

A collision between two continental plates can produce tall mountains. But the mountains form slowly and in stages over millions of years. The history of the Himalayas is illustrated in **Figure 9.** The Himalayas continue to grow even now as continental collision pushes them higher. Note that although the plates move horizontally, the collision causes the crust to move vertically also.

Ocean Trenches

When two plates collide, one can go under the other and be forced into the mantle in a process called subduction. As shown in **Figure 10,** a deep trench forms where the two plates meet. **Ocean trenches** *are deep, underwater troughs created by one plate subducting under another plate at a convergent plate boundary.* Ocean trenches are the deepest places in Earth's oceans.

🔑 **Key Concept Check** What are two landforms that can form where two plates converge?

Volcanic Arcs

Volcanic mountains can form in the ocean where plates converge and one plate subducts under another one. These volcanoes emerge as islands. *A curved line of volcanoes that forms parallel to a plate boundary is called a* **volcanic arc.** Most of the active volcanoes in the United States are part of the Aleutian volcanic arc in Alaska. There are about 40 active volcanoes there. They formed as a result of the Pacific Plate subducting under the North American Plate.

Volcanic arcs in the ocean are also called island arcs. But a volcanic arc can also form where an oceanic plate subducts under a continental plate. Because the continent is above sea level, the volcanoes sit on top of the continent, as does Mount Shasta in California, shown in **Figure 10.**

 Reading Check Where do volcanic arcs form?

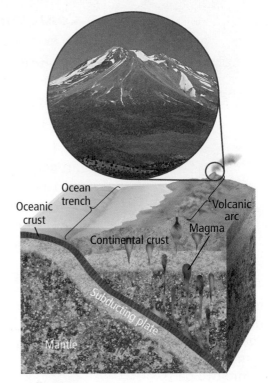

Figure 10 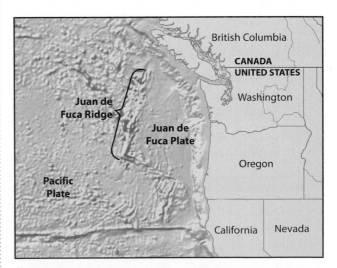 Volcanic arcs also can form on continents. Mount Shasta in California is part of the Cascade volcanic arc.

Landforms Created by Tension

Where plates move apart, tension stresses stretch Earth's crust. Distinct landforms are produced by tension.

Mid-Ocean Ridges

It might surprise you that tension stresses under the ocean can produce long mountain ranges more than 2 km tall. They form under water at divergent boundaries as oceanic plates move away from each other.

As tension stresses cause oceanic crust to spread apart, hot rock from the mantle rises. Because hot rock is less dense than cold rock, the hot mantle pushes the seafloor upward. In this way, long, high ridges are created in Earth's oceans. You might have already learned that a long, tall mountain range that forms where oceanic plates diverge is called a mid-ocean ridge. **Figure 11** shows a mid-ocean ridge near the North American west coast.

Figure 11 The Juan de Fuca Ridge off the coast of Washington and Oregon is a mid-ocean ridge.

Visual Check What direction is the Juan de Fuca Plate moving relative to the Pacific Plate?

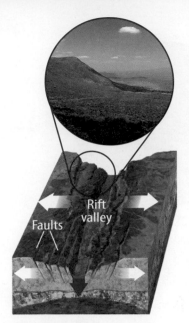

Figure 12 A divergent boundary has created a continental rift in Africa. This rift eventually will separate Africa into two parts.

Continental Rifts

When divergent boundaries occur within a continent, they can form continental rifts, or enormous splits in Earth's crust. Tension stresses in the cold upper part of the crust create faults. At these faults, large blocks of crust move downward, creating valleys between two ridges.

The East African Rift, pictured in **Figure 12,** is an example of an active continental rift that is beginning to split the African continent into two parts. Each year, the two parts move 3–6 mm farther from each other. One day, millions of years from now, the divergent boundary will have created two separate landmasses. Water will fill the space between them.

✓ **Reading Check** Where on Earth is a continental rift forming now?

The valley at this rift also is subsiding. The warm lower part of the crust acts like putty. As the crust stretches, it becomes thinner and subsides, as shown in **Figure 12.**

 Key Concept Check What features form at divergent boundaries?

Inquiry MiniLab

20 minutes

What is the relationship between plate motion and landforms?

As tectonic plates move, they create landforms in predictable patterns. Can you analyze the motion of the plates and predict what landforms will form?

1. Examine the world map shown here indicating the movement of tectonic plates. Determine the meaning of the arrows and the lines.

2. On a **copy of the map,** label plate boundaries as divergent, convergent, or transform.

3. On your map, predict the landforms that will form at each plate boundary.

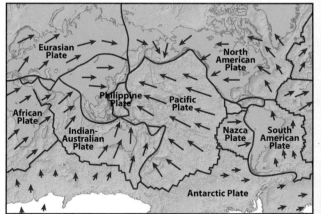

Analyze and Conclude

1. **Describe** the landform that is found around the edge of the Pacific Ocean. Hypothesize why this is called the Ring of Fire.

2. **Compare and contrast** the landforms on the east and west coasts of South America. Support your answer with data.

3. 🔑 **Key Concept** Create a table that relates the type of stress (compression, tension, shear) and the location of the plate boundary (middle of a continent, edge of a continent, middle of an ocean). Fill in the table with the landforms found in each situation. List one place on Earth where each is occurring.

🔑

Figure 13 On the left, a transform fault forms as two plates move past each other. On the right, the yellow line shows the mid-ocean ridge. The red lines are transform faults.

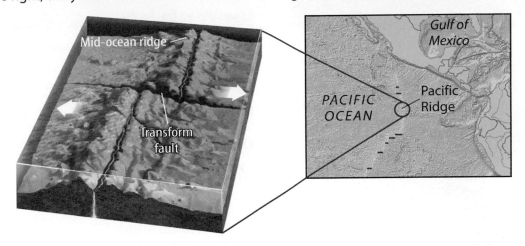

✓ **Visual Check** In the figure on the right, where are the fracture zones?

Landforms Created by Shear Stresses

Recall that as plates slide horizontally past each other, shear stresses produce transform boundaries. Landforms created by shear stresses are not as obvious as landforms created by tension or compression.

Transform Faults

Where tectonic plates slide horizontally past each other, they form **transform faults**. Some transform faults form perpendicular to mid-ocean ridges, as shown on the left in **Figure 13.** Recall that tension produces mid-ocean ridges at divergent boundaries. As the plates slide away from each other, transform faults also form and can separate sections of mid-ocean ridges. The map in **Figure 13** shows the transform faults along the Pacific Ridge.

 Key Concept Check What features form where plates slide past each other?

Fault Zones

Some transform faults can be seen at Earth's surface. For example, the San Andreas Fault in California is visible in many places. Although the San Andreas Fault is visible at Earth's surface, much of this fault system is underground. As shown in **Figure 14,** the San Andreas Fault is not a single fault. Many smaller faults exist in the area around the San Andreas Fault. *An area of many fractured pieces of crust along a large fault is called a* **fault zone.**

WORD ORIGIN ·············

transform
from Latin *trans*, means "across"; and *formare*, means "to form"

Figure 14 Shear stresses create faulting at Earth's surface. Below the surface there might be many other faults that are part of the same fault zone.

🔊

Lesson 2 Review

 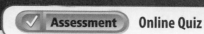

Visual Summary

The deepest and tallest landforms on Earth are created at plate boundaries.

Tension stresses within continents can produce enormous splits in Earth's surface.

Faults at Earth's surface can be part of much larger fault zones that have many underground faults.

FOLDABLES

Use your lesson Foldable to review the lesson. Save your Foldable for the project at the end of the chapter.

What do you think NOW?

You first read the statements below at the beginning of the chapter.

3. New landforms are created only at plate boundaries.

4. The tallest and deepest landforms are created at plate boundaries.

Did you change your mind about whether you agree or disagree with the statements? Rewrite any false statements to make them true.

Use Vocabulary

1 **Use the terms** *volcanic arc* and *trench* in a sentence.

2 **Relate** a transform fault to a fault zone.

3 **Define** *fault zone* in your own words.

Understand Key Concepts

4 Which type of stress is currently producing the East African Rift?
A. shear stress
B. tension stress
C. compression and tension stresses
D. shear and compression stresses

5 **Compare** the development of tall mountains to how ocean trenches form.

6 **Summarize** the processes involved in the formation of a volcanic arc.

Interpret Graphics

7 **Connect** Copy the graphic organizer below. Use it to relate stresses at plate boundaries to the landforms associated with each type of boundary.

Stresses	Landforms
Compression	
Tension	
Shear	

Critical Thinking

8 **Evaluate** If you heard that earthquakes occurred in the same year in Illinois, Missouri, and Kentucky, what could you infer was happening below the surface?

9 **Relate** a continental rift to a mid-ocean ridge.

10 **Create** a diagram that illustrates how mountains form where two continents collide.

Hot Spots!

Volcanoes on a Plate

Not all volcanoes form at plate boundaries. Some, called hot spot volcanoes, pop up in the middle of a tectonic plate. A hot spot volcano forms over a rising column of magma called a mantle plume. The origin of mantle plumes is still uncertain, but evidence shows they probably rise up from the boundary between Earth's mantle and core.

As a tectonic plate passes over a mantle plume, a volcano forms above the plume. The tectonic plate continues to move, and a chain of volcanoes forms. If the volcanoes are in the ocean and if they get large enough, they become islands, such as the Hawaiian Islands. Here is how this happens:

AMERICAN MUSEUM ᴼꜰ NATURAL HISTORY

4 **The oldest islands are farthest from the plume.**

Direction of Pacific Plate motion

Hawaiian Ridge

3 **As the Pacific Plate moves, the islands formed by the hot spot are carried with it and away from the magma plume.**

2 **The seamount continues to grow until it rises above the water and becomes an island.**

Hawaii

1 **Magma, which is less dense than the surrounding rock, rises to the seafloor and forms a seamount.**

Niihau

Kauai
3.8 to 5.6 million years old

Oahu
2.2 to 3.3 million years old

Maui
less than 1.0 million years old

Molokai

Lanai

Kahoolawe

Direction of plate motion

Hawaii
started forming
0.8 million years ago.

It's Your Turn

RESEARCH Not all hot spots arise in oceans. Much of Yellowstone National Park lies inside the caldera of a gigantic volcano that sits on a hot spot. Is Yellowstone's hot spot still active?

Mountain Building

Reading Guide

Key Concepts
ESSENTIAL QUESTIONS

- How do mountains change over time?
- How do different types of mountains form?

Vocabulary
folded mountain p. 271
fault-block mountain p. 272
uplifted mountain p. 273

 Multilingual eGlossary

 Video **BrainPOP®**

inquiry Is this a safe place to live?

Is it safe to live next to this mountain? Will lava erupt from it? Will there be earthquakes nearby? Not all mountains are the same. Once you know how a mountain formed, you can predict what is likely to happen in the future.

What happens when Earth's tectonic plates diverge?

When tectonic plates diverge, the crust gets thinner. Sometimes large blocks of crust subside and form valleys. Blocks next to them move up and become fault-block mountains. This is how the Basin and Range Province in the western United States formed.

1 Stand 5–6 **hardbound books** on a desk with the bindings vertical.

2 Using a **ruler,** measure the width and the height of the books, as shown. Record the results in a table in your Science Journal.

3 Holding the books together, tilt them sideways at the same time to about a 30° angle. Measure the width and the height of the books. Record the results in your table.

4 Tilt the books to about a 60° angle. Measure the width and the height of the books. Record the results in your table.

5 Draw a diagram of the tilted books in your Science Journal, and label mountains, valleys, and faults.

Think About This

1. How does the thickness of the crust relate to the height of a mountain?

2. **Key Concept** How do you think fault-block mountains form?

The Mountain-Building Cycle

Mountain ranges are built slowly, and they change slowly. Because they are the result of many different plate collisions over many millions of years, they are made of many different types of rocks. The processes of weathering and erosion can remove part or all of a mountain.

✓ **Reading Check** What processes can remove part of a mountain?

Converging Plates

Recall that when plates collide at a plate boundary, a combination of folds, faults, and uplift creates mountains. Eventually, after millions of years, the forces that originally caused the plates to move together can become inactive. As shown in **Figure 15,** a single new continent is created from two old ones, and the plate boundary becomes inactive. With no compression at a convergent plate boundary, the mountains stop increasing in size.

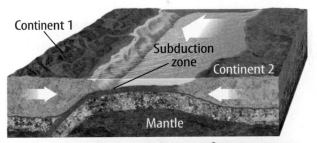

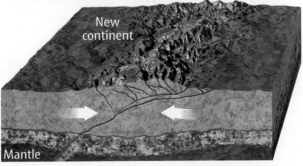

Figure 15 The forces that originally caused plates to move together eventually become inactive. A single continent is created from the two old ones, and the plate boundary becomes inactive.

FOLDABLES®

Fold a sheet of paper to make a vertical three-tab book. Label the tabs as illustrated. Describe how different types of mountains form. Identify a specific example of each type.

Folded Mountain

Fault-block Mountain

Uplifted Mountain

WORD ORIGIN

Appalachian
from the Apalachee *abalahci*, means "other side of the river"

Collisions and Rifting

Continents are continuously changing because Earth's tectonic plates are always moving. When continents split at a divergent plate boundary, they often break close to the place where they first collided. First a large split, or rift, forms. The rift grows, and seawater flows into it, forming an ocean.

Eventually plate motion changes again, and the continents collide. New mountain ranges form on top of or next to older mountain ranges. The cycle of repeated collisions and rifting can create old and complicated mountain ranges, such as the **Appalachian** Mountains.

 Reading Check Where do plates tend to break apart?

Figure 16 illustrates the history of the plate collisions and rifting that produced the mountain range as it is today. Rocks that make up mountain ranges such as the Appalachian Mountains record the history of plate motion and collisions that formed the mountains.

Weathering

The Appalachian Mountains are an old mountain range that stretches along most of the eastern United States. They are not as high and rugged as the Rocky Mountains in the west because they are much older. They are no longer growing. Weathering has rounded the peaks and lowered the elevations.

Formation of the Appalachians  **Concepts in Motion** Animation

Figure 16 The Appalachian Mountains formed over several hundred million years.

Visual Check Which mountain range is between Valley and Ridge and Piedmont?

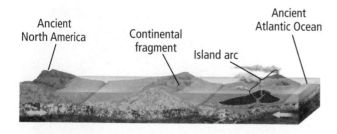

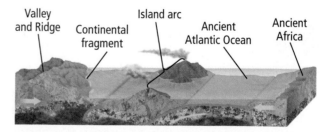

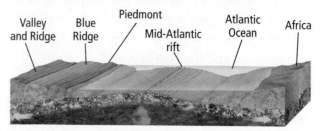

Erosion and Uplift

Over time, natural processes wear down mountains, smooth their peaks, and reduce their height. But some mountain ranges are hundreds of millions of years old. How do they last so long? Recall how isostasy works. As a mountain erodes, the crust under it must rise to restore the balance between what is left of the mountain and how it floats on the mantle. Therefore, rocks deep under continents rise slowly toward Earth's surface. In old mountain ranges, metamorphic rocks that formed deep below the surface are exposed on the top of mountains, such as the rocks in **Figure 17.**

 Key Concept Check How can mountains change over time?

Types of Mountains

You learned in the first lesson that stresses caused by plate movement can pull or compress crust. This is one way plate motion is involved in creating many types of mountains. But the effect of plate movement is also responsible for changing the positions of rocks and the rocks themselves within a mountain range.

Folded Mountains

Rocks that are deeper in the crust are warmer than rocks closer to Earth's surface. Deeper rocks are also under much more pressure. When rocks are hot enough or under enough pressure, folds form instead of faults, as shown in **Figure 18.** **Folded mountains** *are made of layers of rocks that are folded.* Folded mountains form as continental plates collide, folding and uplifting layers of rock. When erosion removes the upper part of the crust, folds are exposed on the surface.

The arrangement of the folds is not accidental. You can demonstrate this by taking a piece of paper and gently pushing the ends toward one another to form a fold. The fold is a long ridge that is **perpendicular** to the direction in which you pushed. Folded mountains are similar. The folds are perpendicular to the direction of the compression that created them.

Figure 17 Metamorphic rocks, such as these, formed deep below Earth's surface. After the material above them eroded, the rock rose due to isostasy. Now they are on Earth's surface.

ACADEMIC VOCABULARY

perpendicular
(adj.) being at right angles to a line or plane

Figure 18 Compression stresses folded these rocks. Because the folds run up and down, the compression must have come from the sides.

Figure 19  In the middle of a continent, tension can pull crust apart. Where the crust breaks, fault-block mountains and valleys can form as huge blocks of Earth rise or fall.

Visual Check Which way is the tension pulling?

Concepts in Motion Animation

Fault-Block Mountains

Sometimes tension stresses within a continent create mountains. As tension pulls crust apart, faults form. At the faults, some blocks of crust fall and others rise, as shown in **Figure 19. Fault-block mountains** *are parallel ridges that form where blocks of crust move up or down along faults.*

The Basin and Range Province in the western United States consists of dozens of parallel fault-block mountains that are oriented north to south. The tension that created the mountains pulled in the east-west directions. One of these mountains is shown at the beginning of this lesson. Notice how a high, craggy ridge is right next to a valley. Somewhere between the two, there is a fault where huge movement once occurred.

Key Concept Check How do folded and fault-block mountains form?

Inquiry **MiniLab**

15 minutes

How do folded mountains form?

When two continental plates converge, rocks crumple and fold, forming folded mountains. If the rocks formed in layers, such as sedimentary rocks, the folds can be visible.

1. Read and complete a lab safety form.

2. On a piece of **waxed paper,** shape four balls of different **colored dough** into rectangles about 1 cm thick.

3. Stack the rectangles on top of each other. Using a **plastic knife,** trim the edges so that all layers are clearly visible. Draw a side view of the unfolded layers in your Science Journal.

4. Compress the dough by pushing the short ends together into an S-shape. Try to get at least one upward and one downward fold. Draw a side view of the folded layers in your Science Journal.

5. Using the knife, simulate erosion by slicing off the top of your folded mountains. Draw a top view of the eroded mountains in your Science Journal.

Analyze and Conclude

1. **Relate** the direction of compression to the direction of the peaks of the mountains.

2. **Key Concept** Describe how folded mountains form and change over time.

Uplifted Mountains

The granite on top of the Sierra Nevada's Mount Whitney was once 10 km below Earth's surface. Now it is on top of a 4,400-m-tall mountain! How did this happen? Mount Whitney is an example of an uplifted mountain. *When large regions rise vertically with very little deformation,* **uplifted mountains** *form.* The rocks in the Sierra Nevada are made of granite, which is an igneous rock originally formed several kilometers below Earth's surface. Uplift and erosion have exposed it.

 Reading Check What type of rocks are found in the Sierra Nevada?

Scientists do not fully understand how uplifted mountains form. One hypothesis proposes that cold mantle under the crust detaches from the crust and sinks deeper into the mantle, as shown in **Figure 20.** The sinking mantle pulls the crust and creates compression closer to the surface. As the crust thickens, the upper part of the crust rises to maintain isostasy. Sometimes it rises high enough to create huge mountain ranges. Geologists are designing experiments to test this hypothesis.

Volcanic Mountains

You might not think of volcanoes as mountains, but scientists consider volcanoes to be special types of mountains. In fact, some of the largest mountains on Earth are made by volcanic eruptions. As molten rock and ash erupt onto Earth's surface, they harden. Over time, many eruptions can build huge volcanic mountains such as the ones that make up the Hawaiian Islands.

Not all volcanic mountains erupt all the time. Some volcanic mountains are dormant, which means they might erupt again someday. Some volcanic mountains will never erupt again.

 Key Concept Check How do uplifted and volcanic mountains form?

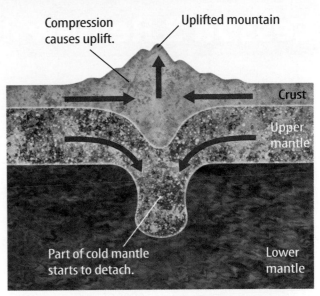

Figure 20 One possible explanation for how uplifted mountains form is that sinking mantle creates compression of the crust. The crust rises to regain isostasy, forming mountains.

Math Skills

Use Proportions

An equation showing two equal ratios is a proportion. Some mountains in the Himalayas are rising 0.001 m/y. How long would it take the mountains to reach a height of 7,000 m?

1. Set up a proportion.

$$\frac{0.001 \text{ m}}{1 \text{ y}} = \frac{7,000 \text{ m}}{x \text{ y}}$$

2. Cross multiply.

$$0.001x = 7,000$$

3. Divide both sides by 0.001.

$$\frac{0.001x}{0.001} = \frac{7,000}{0.001}$$

4. Solve for x.

$$x = 7,000,000 \text{ y}$$

Practice

If the uplift rate of Mount Everest is 0.0006 m/y, how long did it take Mount Everest to reach a height of 8,848 m?

 Review

- **Math Practice**
- **Personal Tutor**

Lesson 3 Review

Visual Summary

Mountain ranges can be the result of repeated continental collision and rifting.

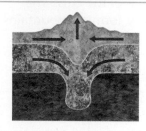

Tension stresses create mountain ranges that are a series of faults, ridges, and valleys.

Uplifted mountains form as a result of compression near Earth's surface.

FOLDABLES

Use your lesson Foldable to review the lesson. Save your Foldable for the project at the end of the chapter.

Use Vocabulary

1 Compression stress can create _____.

2 **Name** two types of mountains that can form far from plate boundaries.

3 Rocks formed deep inside Earth can be found at the surface as _____.

Understand Key Concepts

4 **Contrast** folded and fault-block mountains.

5 Which type of mountains form with little deformation?
  A. fault-block mountains
  B. folded mountains
  C. uplifted mountains
  D. volcanic mountains

6 **Identify** the type of plate boundary where the Appalachian Mountains formed.

Interpret Graphics

7 **Summarize** the plate tectonic events that built the Appalachian Mountains, using a graphic organizer like the one below.

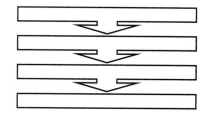

Critical Thinking

8 **Critique** the generalization that mountains only form at convergent boundaries. Explain how other processes can produce mountains.

What do you think NOW?

You first read the statements below at the beginning of the chapter.

5. Metamorphic rocks formed deep below Earth's surface sometimes can be located near the tops of mountains.

6. Mountain ranges can form over long periods of time through repeated collisions between plates.

Did you change your mind about whether you agree or disagree with the statements? Rewrite any false statements to make them true.

Math Skills

Review — Math Practice —

9 Volcanoes in Hawaii began forming on the seafloor, about 5,000 m below the surface. If a volcano reaches the surface in 300,000 years, what was its rate of vertical growth per year?

What tectonic processes are most responsible for shaping North America?

Mountains are important structures of the North American landscape. By studying the types of mountains, scientists can figure out what processes have shaped the continent over the last several hundred million years.

Materials

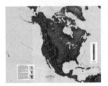

metric ruler

North America map

Learn It

Before scientists can make conclusions about the formation of a continent, they have to know what happened in all of its parts. Scientists **research information** so they can answer questions and draw conclusions about the continent as a whole.

Try It

1 Study the map shown. Choose a mountain range to research. With your teacher's approval, you may research a range that is not shown on this map.

2 Using sources approved by your teacher, research your mountain range. Answer the following questions, and include any other information you find interesting.

- What is the name and location of your mountain range?
- What type of plate boundary is near your mountain range?
- What tectonic plates form the boundary near your mountain range?
- What type(s) of mountains make up your mountain range?
- How did the mountains form?
- What type(s) of rocks make up the mountains?
- How tall are the mountains in your range?
- How old is your mountain range?
- What other factors have affected the height or the shape of the mountains?

3 Record your results in your Science Journal.

Apply It

4 **Create** a visual presentation about the mountain range you researched. Include photographs or sketches to support your research.

5 **Compare and Contrast** Combine your research with the other groups' research. How are your mountain ranges similar? How are they different?

6 **Key Concept** What forces have shaped North America?

Continent Building

Key Concepts 🔑
ESSENTIAL QUESTIONS

- What are two ways continents grow?
- What are the differences between interior plains, basins, and plateaus?

Vocabulary

plains p. 279
basin p. 279
plateau p. 280

🄖 **Multilingual eGlossary**

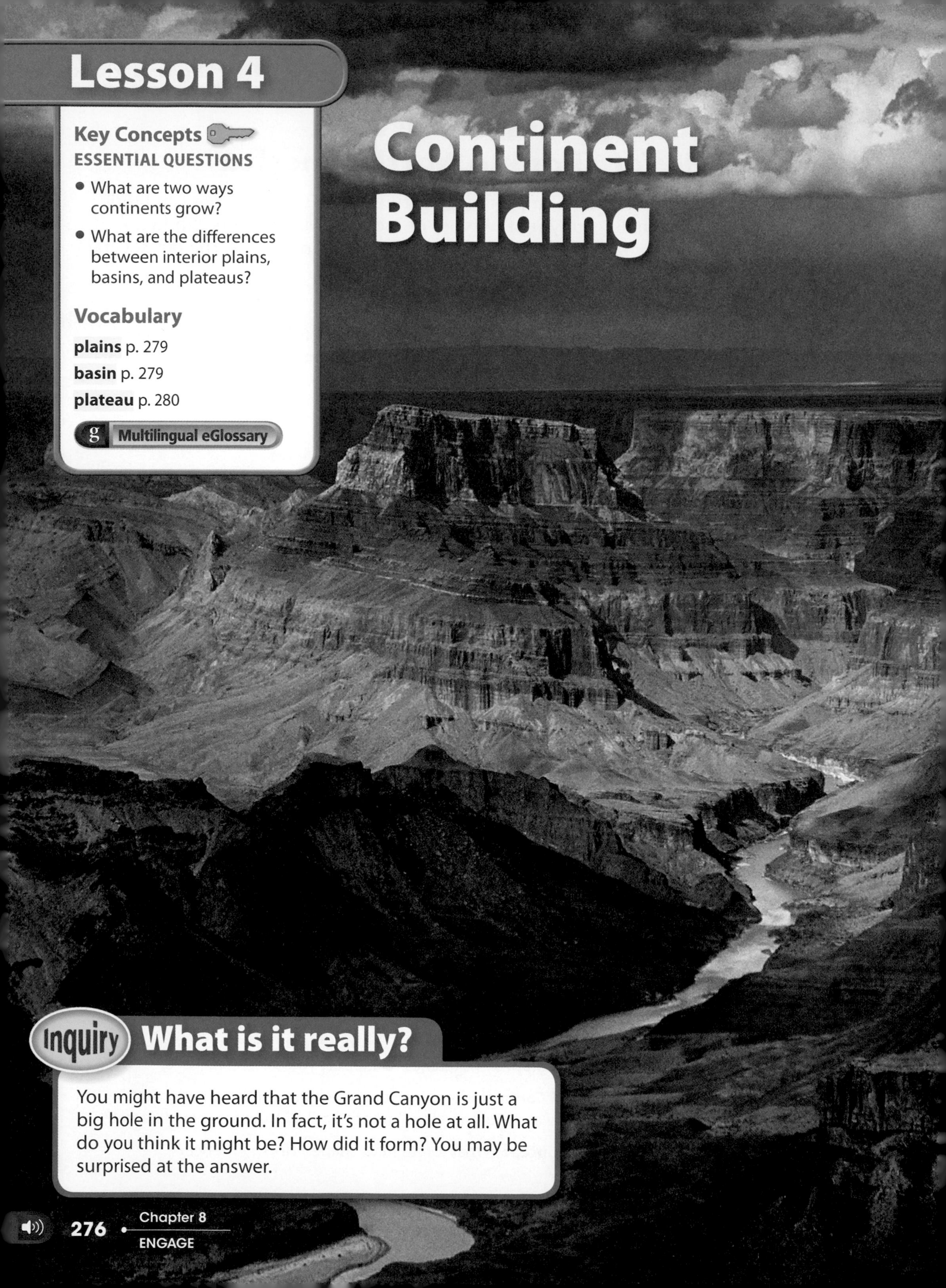

Inquiry What is it really?

You might have heard that the Grand Canyon is just a big hole in the ground. In fact, it's not a hole at all. What do you think it might be? How did it form? You may be surprised at the answer.

How do continents grow?

Over the history of Earth, continents have been slowly increasing in size. Continents can grow when fragments of crust that formed in other parts of the world stick to the edges of the continent at convergent plate boundaries.

1. Read and complete a lab safety form.
2. Place **waxed paper** on the lab bench, and place a **block of wood** on one end of the paper.
3. Using **shaving cream,** create a volcanic arc on the waxed paper.
4. Pull the waxed paper under the wood and observe what happens. Record your observations in your Science Journal.

Think About This

1. Using the vocabulary from the chapter (words such as *compression, convergence, folded mountains, volcanic mountains*), describe what occurred as you completed the lab.

2. Create a labeled diagram showing the motion of the ocean plate and the continental plate. Include the volcanic arc and describe what happened to it when it ran into the continent.

3. 🔑 **Key Concept** How do you think continents grow?

The Structure of Continents

If you look at the map shown in **Figure 21,** you will notice that most of the highest elevations are located near the edges of continents. Why do you think that is?

In contrast, the interiors of most continents are flat. Usually, the middle of a continent is only a few hundred meters above sea level. Continental interiors have very few mountains. In these regions, the rocks exposed at Earth's surface are old igneous and metamorphic rocks. A map showing the old, stable interiors of the world's continents is on the right in **Figure 21.** Notice that they usually lie near the middle of the continent. These areas are usually smooth and flat because millions or even billions of years of erosion have smoothed them out.

 Reading Check Where are high elevations usually located? Where are the low elevations?

WORD ORIGIN ············

continent
from Latin *terra continens,*
means "continuous land"

Figure 21 The map on the left shows areas of high elevation in white. They are usually near the edges of continents. The map on the right shows continental interiors that have low elevation. 🔑

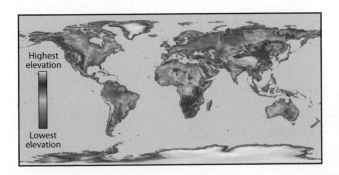

Highest elevation

Lowest elevation

How Continents Grow

The shapes and the sizes of the continents have changed many times over Earth's history. Continents can break up and get smaller, or they can get bigger. One way continents get bigger is through the addition of igneous rocks by erupting volcanoes. A second way is when tectonic plates carry island arcs, whole continents, or fragments of continents with them.

When a plate carrying fragments reaches a continent at a convergent boundary, the least dense fragments get pushed onto the edge of the continent. **Figure 22** is a map showing fragments that have been added to the west coast of North America within the last 600 million years. Fragments in the western United States include volcanic arcs, ancient seafloor, and small pieces from other continents.

Figure 22 The green areas show parts of present-day North America that were once attached to other continents in other parts of the world.

Animation

✓ **Key Concept Check** What are two ways continents grow?

If rock were not being added to continents, what do you think would happen to the size of continents? Rifting would change their sizes and shapes. Weathering and erosion gradually would wear them down.

Inquiry MiniLab
20 minutes

Can you analyze a continent?

As tectonic plates move across Earth's surface, they interact in predictable patterns. Suppose you could get a glimpse of a continent on Earth at some other time. You can use what you know to figure out what that continent is like.

1. Read and complete a lab safety form.

2. Copy the imaginary continent shown at right into your Science Journal. Arrow length is proportional to the speed of the plates.

3. Use **colored pencils** to differentiate regions of compression, tension, and shear.

4. Identify the locations and types of landforms that would be present on Gigantia. Label fault-block mountains, faults, and folded mountains.

5. Determine the locations of the interior plains and where continental fragments are being added.

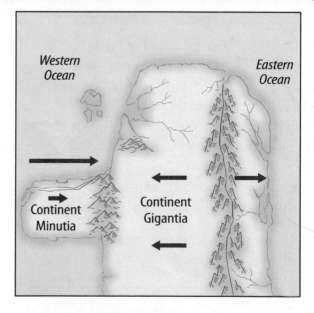

Analyze and Conclude

1. **Select** a region of Earth that has similar plate interactions to those on Gigantia.

2. 🔑 **Key Concept** Describe how the continent is changing.

Continental Interiors

Rocks in continental interiors tend to be stable, flat, very old, and very strong. They are usually more than 500 million years old. In some continental interiors, the rocks are much older than that! **Figure 23** shows rocks that might be the oldest on Earth's surface. Although they might not look very exciting, it is incredible to think they are more than 4.2 billion years old!

Formation of Interior Plains

A **plain** is *an extensive area of level or rolling land.* Most of the central region of North America is referred to as the Interior Plains. The rocks in these plains came from collisions of several smaller plates about 1 billion years ago. At different times in Earth's history, the plains were covered by shallow seas. The plains have been flattened by millions of years of weathering and erosion.

 Key Concept Check What is a plain?

Formation of Basins

Just as plate motion and isostasy create mountains, they also can cause subsidence. *Areas of subsidence and regions with low elevation are called* **basins.** Sediments eroded from mountains accumulate in basins. **Figure 24** is a map showing the largest basins in North America. Can you find a relationship between the locations of basins and large mountain ranges?

 Reading Check What is the name of the feature where sediment accumulates?

Basins can have great economic importance. Under the right conditions, the remains of plants and animals are buried in the sediments that accumulate in basins. Over millions of years, heat and pressure convert the plant and animal remains into oil, natural gas, and coal. Most of our energy resources are extracted from sedimentary basins. The world's largest oil and natural gas fields also lie in sedimentary basins.

Figure 23 The Canadian rocks pictured here have existed throughout much of Earth's history. They are more than 4.2 billion years old.

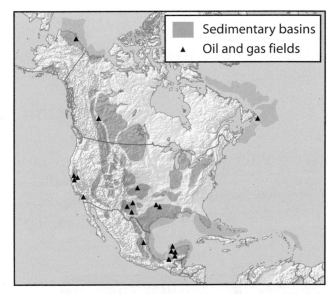

Figure 24 Ancient sedimentary basins are important because oil, natural gas, and coal usually are found in basins.

Visual Check Where are oil and gas fields in relation to sedimentary basins?

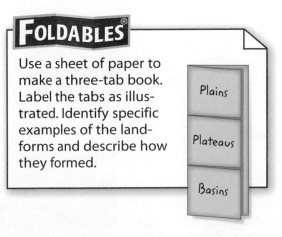

FOLDABLES

Use a sheet of paper to make a three-tab book. Label the tabs as illustrated. Identify specific examples of the landforms and describe how they formed.

Plains

Plateaus

Basins

Figure 25 The Colorado Plateau is an example of an uplifted plateau.

🔍 **Visual Check** Which states are partly covered by the Colorado Plateau?

Formation of Plateaus

Some regions are high above sea level but are flat. *Flat regions with high elevations are called* **plateaus.** Some plateaus form through uplift. An example of an uplifted plateau is the Colorado Plateau, shown in **Figure 25.** In the last 5 million years, this region has been uplifted by more than 1 km.

Notice in **Figure 25** that the Grand Canyon is only a small part of the Colorado Plateau. It was created as the Colorado River cut through and eroded the uplifting plateau. So, the Grand Canyon was created by water!

The eruption of lava also can create large plateaus. More than 200,000 km³ of lava flooded the large area shown in **Figure 26.** Over 2 million years, multiple eruptions built up layers of rock. In some places, the plateau shown in the photo in **Figure 26** is more than 3 km thick!

🔑 **Key Concept Check** What are the differences between plains, basins, and plateaus?

Dynamic Landforms

When you started reading this chapter, you might have thought Earth had always looked the same. Now you know that Earth's surface is constantly changing. Mountains form only to be eroded away. Continents grow, shift, and shrink. Nothing stays the same for long on dynamic Earth.

The Columbia Plateau 🔑

Figure 26 The map shows the area that was covered by multiple eruptions of lava over millions of years. The lava cooled and formed the Columbia River basalt. The layers of basalt are visible in the photograph. Some parts of the Columbia Plateau are more than 3 km thick.

Visual Summary

Rocks at the center of most continents are very old, very strong, and flat.

Fragments of crust are added to continents at convergent boundaries.

Large, elevated plateaus are created through uplift and lava flows.

FOLDABLES®

Use your lesson Foldable to review the lesson. Save your Foldable for the project at the end of the chapter.

What do you think NOW?

You first read the statements below at the beginning of the chapter.

7. The centers of continents are flat and old.

8. Continents are continually shrinking because of erosion.

Did you change your mind about whether you agree or disagree with the statements? Rewrite any false statements to make them true.

Use Vocabulary

1 As a mountain erodes, sediment can accumulate in a nearby _____.

2 The central, flat region of North America is known as the _____.

3 The Grand Canyon was eroded out of a large _____.

Understand Key Concepts

4 Which term best describes the center of North America?
- **A.** basin
- **B.** lava
- **C.** plateau
- **D.** plain

5 **Describe** how continents change over time.

6 **Contrast** basins and plateaus.

Interpret Graphics

7 **Summarize** Use a graphic organizer like the one below to show the different stages involved in continent growth. Begin with an old continental interior and end with a new continent.

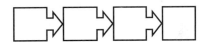

Critical Thinking

8 **Infer** In this lesson you learned that fragments of other continents were added to the west coast of North America. Where in the United States have other continental fragments been added?

9 **Generalize** How are landforms near the edges of continents different from landforms in continental interiors? How are these landforms related to plate tectonics or processes in the rock cycle?

10 **Infer** What would happen to the Grand Canyon if there were further uplift of the Colorado Plateau?

Design Landforms

Suppose you are a museum designer and you want to show people that different landforms form under different circumstances. Sometimes rocks fold, sometimes they break, sometimes they form mountains, and sometimes they sink into Earth and create trenches. What they do depends on the properties of the rock and the type of stress. Unfortunately, rocks do all of these things so slowly that it is hard to see rocks in motion. What materials would you use to model the formation of landforms? What factors affect how rocks behave? How could you change your materials to model how rocks behave?

Ask a Question

What materials could represent rocks? How are the materials different from rocks?

Make Observations

1. Read and complete the lab safety form.

2. Mix some ingredients in your plastic bin or mixing bowl. Try different combinations until you make a material you can use to model landforms.

3. Experiment with the materials, and try to create different landforms.

4. Record your observations in your Science Journal. How do the materials behave like rocks, and how are they different?

Form a Hypothesis

5 After observing the behavior of your material, think of factors that cause rocks to behave differently. How might you recreate these different situations? Pick one factor and develop a hypothesis about how you can use the materials to model the behavior of rocks.

Test Your Hypothesis

6 Develop a procedure to test your hypothesis. What is your dependent variable and your independent variable? How will you make quantitative measurements of both variables?

7 Create a table to record your results.

8 Have your teacher approve your procedure and your table.

9 Conduct your experiment, and record your results.

Analyze and Conclude

10 **Create** a graph displaying your results.

11 **Interpret** your graph and explain the relationship between the variables.

12 **Critique** your procedure and your results.

13 **The Big Idea** Relate your results to how Earth's surface is shaped by plate motion.

Communicate Your Results

Design a museum exhibit that models the formation of one or more landforms.

Inquiry Extension

What materials could represent Earth's crust? Now that you have modeled Earth's landforms, put the landforms on tectonic plates. Model plate motion, and describe how your landforms change at different types of plate boundaries.

Lab Tips

☑ This lab might be messy! Clean up dry cornstarch and flour with a broom and dustpan.

☑ Be quantitative! Figure out concrete ways to measure both the variable you are changing (the independent variable) and the variable you are measuring (the dependent variable).

☑ If your first try is not successful, try something different! Science rarely works on the first try.

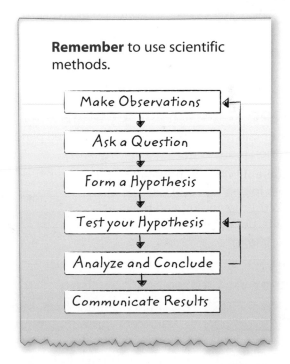

Remember to use scientific methods.

Make Observations

Ask a Question

Form a Hypothesis

Test your Hypothesis

Analyze and Conclude

Communicate Results

The forces created by the movement of tectonic plates are responsible for the variety of Earth's constantly changing landforms.

Key Concepts Summary 🔑	Vocabulary
Lesson 1: Forces That Shape Earth • As continents float in the mantle, they rise and fall to maintain the balance of **isostasy.** • **Compression, tension,** and **shear** stress can deform or break rocks. • **Uplift** and plate motion move rocks through the rock cycle. 	**isostasy** p. 254 **subsidence** p. 255 **uplift** p. 255 **compression** p. 255 **tension** p. 255 **shear** p. 255 **strain** p. 256
Lesson 2: Landforms at Plate Boundaries  • When two continental plates collide, tall mountain ranges form. When an oceanic plate subducts below another one, an **ocean trench** and a **volcanic arc** form. • At divergent boundaries, mid-ocean ridges and continental rifts form. • **Transform faults** can create large areas of faulting and fracturing, not all of which can be seen at Earth's surface.	**ocean trench** p. 262 **volcanic arc** p. 263 **transform fault** p. 265 **fault zone** p. 265
Lesson 3: Mountain Building • Mountain ranges can grow from repeated plate collisions. Erosion reduces the sizes of continents. • Different types of mountains form from folded layers of rock, blocks of crust moving up and down at faults, uplift, and volcanic eruptions. 	**folded mountain** p. 271 **fault-block mountain** p. 272 **uplifted mountain** p. 273
Lesson 4: Continent Building  • Continents shrink because of erosion and rifting. Continents grow through volcanic activity and continental collisions. • **Plains** are generally flat areas of land, usually in the center of continents. **Basins** are regions at low elevation where sediment accumulates or once accumulated. **Plateaus** are large, flat regions at high elevation.	**plains** p. 279 **basin** p. 279 **plateau** p. 280

▣ Review
• **Personal Tutor**
• **Vocabulary eGames**
• **Vocabulary eFlashcards**

FOLDABLES® **Chapter Project**

Assemble your lesson Foldables as shown to make a Chapter Project. Use the project to review what you have learned in this chapter.

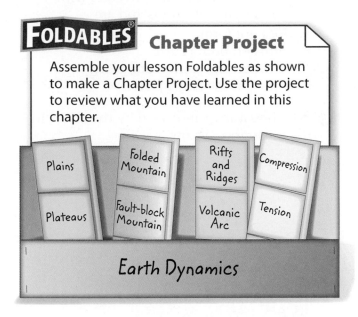

Plains

Folded Mountain

Rifts and Ridges

Compression

Plateaus

Fault-block Mountain

Volcanic Arc

Tension

Earth Dynamics

Use Vocabulary

1 Plastic and elastic deformation are types of _____.

2 Areas of fractured crust along a fault are called _____.

3 Mountains that rise with little deformation of rock are _____.

4 Repeated volcanic eruptions on land can create large _____.

5 The downward vertical motion of Earth's surface is _____.

6 Parallel ridges separated by faults and valleys are _____.

Link Vocabulary and Key Concepts

(((○ **Concepts in Motion** **Interactive Concept Map**

Copy this concept map, and then use vocabulary terms from the previous page and other terms from the chapter to complete the concept map.

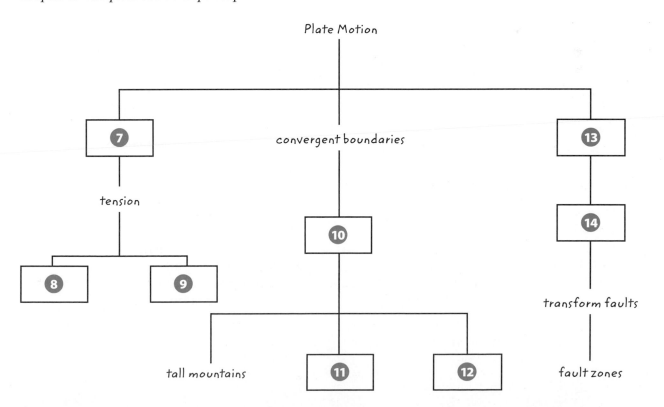

Plate Motion

7

convergent boundaries

13

tension

10

14

8 **9**

transform faults

tall mountains

11 **12**

fault zones

Understand Key Concepts

1 The fact that land surface is high where crust is thick is due to what?

A. isostasy
B. subduction
C. shear stresses
D. tension stresses

2 The highest mountains form at which type of plate boundary?

A. convergent
B. divergent
C. oceanic
D. transform

3 Why does plastic deformation occur in the lower crust?

A. Rocks are hot.
B. Rocks are strong.
C. Tension occurs in the lower crust.
D. The mantle is plastic.

4 Lake Baikal in Siberia, pictured below, fills a continental rift valley. What type of stress is creating the rift?

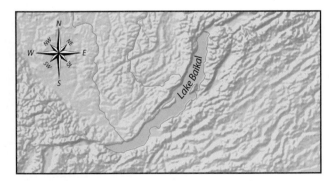

A. compression in the north-south direction
B. shear in the northeast-southwest direction.
C. tension in the north-south direction
D. tension in the northwest-southeast direction

5 When an oceanic plate converges with a continental plate, an island arc

A. does not form.
B. forms on both plates.
C. forms on the continental plate.
D. forms on the larger plate.

6 Which feature is indicated by the arrow in the illustration below?

A. a fault zone
B. an ocean trench
C. an uplifted mountain
D. a volcanic arc

7 Where in the United States have continental fragments been added?

A. in the center
B. on the east and west coasts
C. on the east coast
D. only close to the Gulf of Mexico

8 Which are the cause of rocks being exposed at Earth's surface?

A. erosion and subsidence
B. erosion and uplift
C. faulting and folding
D. folding and subsidence

Critical Thinking

9 Compare a floating iceberg with a continent floating on the mantle.

10 Explain how seashells got on top of Mount Everest.

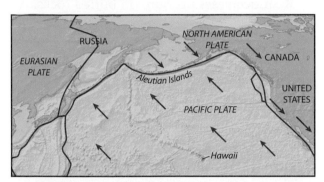

11 Infer Look at the illustration above. Where will the Hawaiian Islands be added to another continent?

12 Assess the statement, "Isostasy never stops causing uplift and subsidence."

13 Suggest the source for the sediment that filled the basins in North Dakota and Colorado.

14 Defend the statement that the large mountains of the Hawaiian Islands did not form at a plate boundary.

15 Predict where on Earth the crust is thickest.

16 Infer Why is the area in southern Africa flat but elevated?

17 Illustrate a possible future for the Appalachian Mountains. How do you think the Appalachian Mountains will look in 200 million years? What processes will change these mountains?

Writing in Science

18 Write a paragraph describing the history of an imaginary mountain range. Use the terms *fold, volcanic arc, convergent,* and *divergent* in your paragraph.

REVIEW THE BIG IDEA

19 If plate tectonics were suddenly to stop, how would Earth's surface change?

20 The photo below shows Mount Everest in the Himalayas. How did it get to be so tall?

Math Skills ×÷

Review — Math Practice

Use Proportions

21 The Himalayas formed when the Indian sub-continent collided with the Eurasian Plate. The Indian subcontinent moved about 10 cm/y.

a. How far would it have moved in 24,000,000 years?

b. How many kilometers did the plate move? (1 km = 100,000 cm)

22 A continent travels 0.006 m/y. How long would it take the continent to travel 100 m?

23 Mount Whitney is 4,419 m high. It started as a hill with an elevation of only 457 meters about 40 mya. What was the rate of uplift for Mount Whitney in m/y? (Hint: Figure out the total elevation gain first.)

Standardized Test Practice

Record your answers on the answer sheet provided by your teacher or on a sheet of paper.

Multiple Choice

1 Which is the result of isostasy?

   A a basin filling with sediment

   B an iceberg floating in the ocean

   C magma rising beneath a mountain

   D one plate subducting under another one

2 The San Andreas Fault is classified as a transform fault. Which type of stress can create transform faults?

   A compression

   B shear

   C fracture

   D tension

Use the figure below to answer question 3.

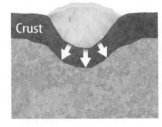

3 Which process is shown in the figure?

   A compression

   B shearing

   C subsidence

   D uplifting

4 What happens when rock fails?

   A It breaks.

   B It deforms elastically.

   C It deforms plastically.

   D It folds.

5 What part does subduction play in the rock cycle?

   A It breaks rocks into sediment.

   B It decreases pressure on buried rocks.

   C It pulls rocks deep into Earth.

   D It pushes rocks up to Earth's surface.

Use the figure below to answer questions 6 and 7.

6 Which force is shown in the figure?

   A compression

   B shearing

   C tension

   D uplift

7 What type of mountain results from the force shown in the figure?

   A folded

   B fault-block

   C uplifted

   D volcanic

8 Which continental landform can result when two plates diverge?

   A basin

   B continental rift

   C mid-ocean ridge

   D transform fault

Use the figure below to answer question 9.

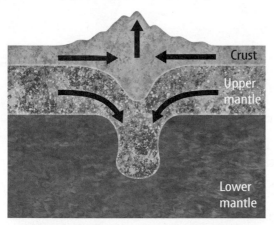

9 Which kind of mountain is shown in the figure?

A fault-block

B folded

C uplifted

D volcanic

10 Which feature is a flat region at a high elevation?

A a basin

B a mountain

C a plain

D a plateau

11 Which landforms are most likely to have coal, oil, and gas deposits?

A basins

B mountains

C plains

D plateaus

Constructed Response

Use the figure below to answer questions 12 and 13.

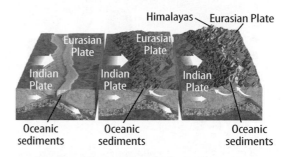

12 Use the figure to explain how the Himalayas formed. Identify the forces involved.

13 What would happen if the plate motion were reversed? Describe possible scenarios for the stages shown on the left and the right of the figure.

Use the figure below to answer question 14.

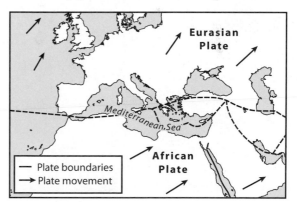

14 Each year, the African Plate moves closer to the Eurasian Plate. Predict how the Mediterranean Sea and the continents, shown in the figure, will change in 100 million years.

NEED EXTRA HELP?														
If You Missed Question...	1	2	3	4	5	6	7	8	9	10	11	12	13	14
Go to Lesson...	1	2	1	2	1	1	2	3	4	4	2	3	4	4

Chapter 9

Earthquakes and Volcanoes

THE BIG IDEA

What causes earthquakes and volcanic eruptions?

Inquiry **Why do volcanoes erupt?**

Mount Pinatubo, a volcano in the Philippines, ejected superheated particles of ash and dust in June 1991. This truck is trying to outrun a pyroclastic flow produced during this eruption. *Pyroclastic* means "fire fragments." Why do you suppose this eruption was so dangerous?

- Why did Mount Pinatubo erupt explosively?

- Can scientists predict earthquakes and volcanic eruptions?

- What causes earthquakes and volcanic activity?

Get Ready to Read

What do you think?

Before you read, decide if you agree or disagree with each of these statements. As you read this chapter, see if you change your mind about any of the statements.

1 Earth's crust is broken into rigid slabs of rock that move, causing earthquakes and volcanic eruptions.

2 Earthquakes create energy waves that travel through Earth.

3 All earthquakes occur on plate boundaries.

4 Volcanoes can erupt anywhere on Earth.

5 Volcanic eruptions are rare.

6 Volcanic eruptions only affect people and places close to the volcano.

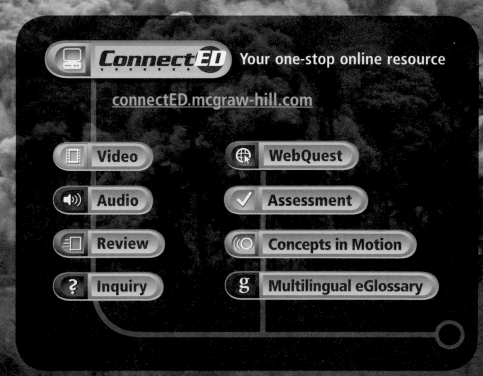

ConnectED Your one-stop online resource

connectED.mcgraw-hill.com

- Video
- WebQuest
- Audio
- Assessment
- Review
- Concepts in Motion
- ? Inquiry
- g Multilingual eGlossary

Earthquakes

Inquiry **Why did this building collapse?**

This building collapsed during the Loma Prieta earthquake that shook the San Francisco Bay area of California in 1989. The magnitude 7.1 earthquake produced severe shaking and damage. Freeways and buildings collapsed and a number of injuries and fatalities occurred. Why are earthquakes common in California?

What causes earthquakes?

Earthquakes occur every day. On average, approximately 35 earthquakes happen on Earth every day. These earthquakes vary in severity. What causes the intense shaking of an earthquake? In this activity, you will simulate the energy released during an earthquake and observe the shaking that results.

1. Read and complete a lab safety form.
2. Tie two **large, thick rubber bands** together.
3. Loop one rubber band lengthwise around a **textbook.**
4. Use **tape** to secure a sheet of **medium-grained sandpaper** to the tabletop.
5. Tape a second sheet of sandpaper to the cover of the textbook.
6. Place the book on the table so that the sheets of sandpaper touch.
7. Slowly pull on the end of the rubber band until the book moves.
8. Observe and record what happens in your Science Journal.

Think About This

1. How does this experiment model the buildup of stress along a fault?

2. **Key Concept** Why does the rapid movement of rocks along a fault result in an earthquake?

What are earthquakes?

Have you ever tried to bend a stick until it breaks? When the stick snaps, it vibrates, releasing energy. Earthquakes happen in a similar way. **Earthquakes** *are the vibrations in the ground that result from movement along breaks in Earth's lithosphere.* These breaks are called faults.

Key Concept Check What is an earthquake?

Why do rocks move along a fault? The forces that move tectonic plates also push and pull on rocks along the fault. If these forces become large enough, the blocks of rock on either side of the fault can move horizontally or vertically past each other. The greater the force applied to a fault, the greater the chance of a large and destructive earthquake. **Figure 1** shows earthquake damage from the Northridge earthquake in 1994.

Figure 1 In 1994, the Northridge earthquake along the San Andreas Fault in California caused $20 billion in damage.

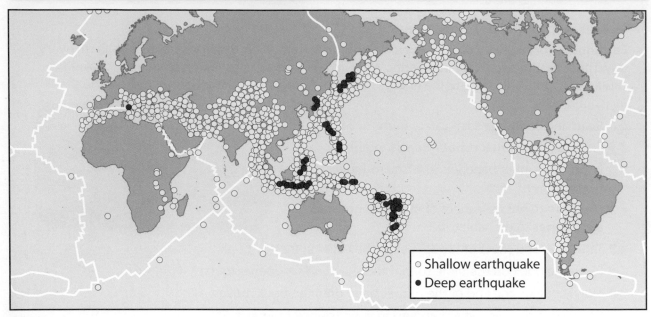

○ Shallow earthquake
● Deep earthquake

Figure 2 Notice that most earthquakes occur along plate boundaries.

Where do earthquakes occur?

The locations of major earthquakes that occurred between 2000 and 2008 are shown in **Figure 2.** Notice that only a few earthquakes occurred in the middle of a continent. Records show that most earthquakes occur in the oceans and along the edges of continents. Are there any exceptions?

Earthquakes and Plate Boundaries

Compare the location of earthquakes in **Figure 2** with tectonic **plate boundaries.** What is the relationship between earthquakes and plate boundaries? Earthquakes result from the buildup and release of stress along active plate boundaries.

Some earthquakes occur more than 100 km below Earth's surface, as shown in **Figure 2.** Which plate boundaries are associated with deep earthquakes? The deepest earthquakes occur where plates collide along a convergent plate boundary. Here, the denser oceanic plate subducts into the **mantle.** Earthquakes that occur along convergent plate boundaries typically release tremendous amounts of energy. They can also be disastrous.

Shallow earthquakes are common where plates separate along a divergent plate boundary, like the mid-ocean ridge system. Shallow earthquakes can also occur along transform plate boundaries like the San Andreas Fault in California. Earthquakes of varying depths occur where continents collide. Continental collisions result in the formation of large and deformed mountain ranges such as the Himalayas in Asia.

✓ **Key Concept Check** Where do most earthquakes occur?

Rock Deformation

At the beginning of this lesson, you read that earthquake energy is similar to bending and breaking a stick. Rocks below Earth's surface behave the same way. When a force is applied to a body of rock, depending on the properties of the rock and the force applied, the rock might bend or break.

When a force such as pressure is applied to rock along plate boundaries, the rock can change shape. This is called rock deformation. Eventually the rocks can be deformed so much that they break and move. **Figure 3** illustrates how rock deformation can result in ground displacement. Notice that rock deformation has resulted in ground displacement where the creek has been pulled in two different directions.

Faults

When stress builds in places like a plate boundary, rocks can form faults. *A **fault** is a break in Earth's lithosphere where one block of rock moves toward, away from, or past another.* When rocks move in any direction along a fault, an earthquake occurs. The direction that rocks move on either side of the fault depends on the forces applied to the fault. **Table 1** lists three types of faults that result from motion along plate boundaries. These faults are called strike-slip, normal, and reverse faults.

 Reading Check What is a fault?

▲ **Figure 3** Forces at work along the San Andreas Fault in California caused displacement of this creek in two directions along a strike-slip fault.

Table 1 The three types of faults are defined based on relative motion along the fault. ▼

Table 1 Types of Faults		
Strike-slip	• Two blocks of rock slide horizontally past each other in opposite directions. • Location: transform plate boundaries	
Normal	• Forces pull two blocks of rock apart. The block of rock above the fault moves down relative to the block of rock below the fault. • Location: divergent plate boundaries	
Reverse	• Forces push two blocks of rock together. The block of rock above the fault moves up relative to the block of rock below the fault. • Location: convergent plate boundaries	

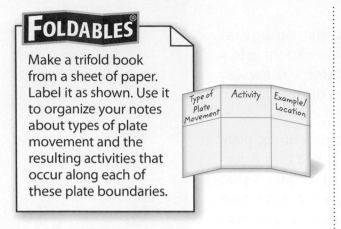
Types of Faults Strike-slip faults can form along transform plate boundaries. There, forces cause rocks to slide horizontally past each other in opposite directions. In contrast, normal faults can form when forces pull rocks apart along a divergent plate boundary. At a normal fault, one block of rock moves down relative to the other. Forces push rocks toward each other at a convergent plate boundary and a reverse fault can form. There, one block of rock moves up relative to another block of rock.

Reading Check What are the three types of faults?

Earthquake Focus and Epicenter

When rocks move along a fault, they release *energy that travels as vibrations on and in Earth called* **seismic waves.** *These waves originate where rocks first move along the fault, at a location inside Earth called the* **focus.** Earthquakes can occur anywhere between Earth's surface and depths of greater than 600 km. When you watch a news report, the reporter often will identify the earthquake's epicenter. *The* **epicenter** *is the location on Earth's surface directly above the earthquake's focus.* **Figure 4** shows the relationship between an earthquake's focus and its epicenter.

SCIENCE USE V. COMMON USE · · · · · · · · · · · · · ·

focus

Science Use the place of origin of an earthquake

Common Use to concentrate

· ·

Figure 4 An earthquake epicenter is above a focus, where the motion along the fault first occurs.

Visual Check What is the relationship between an earthquake focus and an epicenter?

Seismic Waves

During an earthquake, a rapid release of energy along a fault produces seismic waves. Seismic waves travel outward in all directions through rock. It is similar to what happens when you drop a stone into water. When the stone strikes the water's surface, ripples move outward in circles. Seismic waves transfer energy through the ground and produce the motion that you feel during an earthquake. The energy released is strongest near the epicenter. As seismic waves move away from the epicenter, they decrease in energy and intensity. The farther you are from an earthquake's epicenter, the less the ground moves.

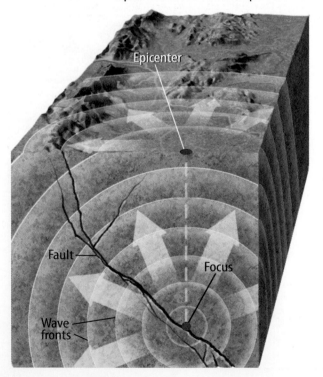

Epicenter

Fault

Focus

Wave fronts

Types of Seismic Waves

When an earthquake occurs, particles in the ground can move back and forth, up and down, or in an elliptical motion parallel to the direction the seismic wave travels. Scientists use wave motion, wave speed, and the type of material that the waves travel through to classify seismic waves. The three types of seismic waves are **primary** waves, secondary waves, and surface waves.

As shown in **Table 2, primary waves,** *also called P-waves, cause particles in the ground to move in a push-pull motion similar to a coiled spring.* P-waves are the fastest-moving seismic waves. They are the first waves that you feel following an earthquake. **Secondary waves,** *also called S-waves, are slower than P-waves. They cause particles to move up and down at right angles relative to the direction the wave travels.* This movement can be demonstrated by shaking a coiled spring side to side and up and down at the same time. **Surface waves** *cause particles in the ground to move up and down in a rolling motion,* similar to ocean waves. Surface waves travel only on Earth's surface closest to the epicenter. P-waves and S-waves can travel through Earth's interior. However, scientists have discovered that S-waves cannot travel through liquid.

 Reading Check Describe the three types of seismic waves.

WORD ORIGIN · · · · · · · · · · ·

primary
from Latin *primus,* means "first"

Table 2 The three types of seismic waves are classified by wave motion, wave speed, and the types of materials they can travel through.

Concepts in Motion
Animation

Table 2 Properties of Seismic Waves

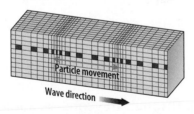
Particle movement
Wave direction

Primary wave
- Cause rock particles to vibrate in the same direction that waves travel
- Fastest seismic waves
- First to be detected and recorded
- Travel through solids and liquids

Secondary wave
- Cause rock particles to vibrate perpendicular to the direction that waves travel
- Slower than P-waves, faster than surface waves
- Detected and recorded after P-waves
- Only travel through solids

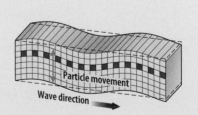
Particle movement
Wave direction

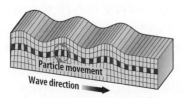
Particle movement
Wave direction

Surface wave
- Cause rock particles to move in a rolling or elliptical motion in the same direction that waves travel
- Slowest seismic wave
- Generally cause the most damage at Earth's surface

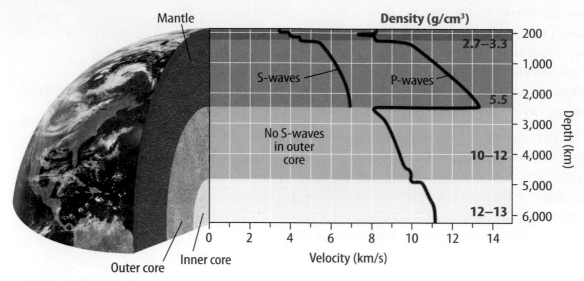

Mantle

Density (g/cm³)

2.7–3.3

S-waves   P-waves

5.5

No S-waves
in outer
core

10–12

12–13

Depth (km)

200
1,000
2,000
3,000
4,000
5,000
6,000

0   2   4   6   8   10   12   14

Velocity (km/s)

Outer core   Inner core

Figure 5 Seismic waves change speed and direction as they travel through Earth's interior. S-waves do not travel through Earth's outer core because it is liquid.

 Visual Check What happens to P-waves and S-waves at a depth of 2500 km?

Mapping Earth's Interior

Scientists that study earthquakes are called **seismologists** (size MAH luh just). They use the properties of seismic waves to map Earth's interior. P-waves and S-waves change speed and direction depending on the material they travel through. **Figure 5** shows the speed of P-waves and S-waves at different depths within Earth's interior. By comparing these measurements to the densities of different Earth materials, scientists have determined the composition of Earth's layers.

Inner and Outer Core Through extensive earthquake studies, seismologists have discovered that S-waves cannot travel through the outer core. This discovery proved that Earth's outer core is liquid unlike the solid inner core. By analyzing the speed of P-waves traveling through the core, seismologists also have discovered that the inner and outer cores are composed of mostly iron and nickel.

✓ **Reading Check** How did scientists discover that Earth's outer core is liquid?

The Mantle Seismologists also have used seismic waves to model convection currents in the mantle. The speeds of seismic waves depend on the temperature, pressure, and chemistry of the rocks that the seismic waves travel through. Seismic waves tend to slow down as they travel through hot material. For example, seismic waves are slower in areas of the mantle beneath mid-ocean ridges or near hotspots. Seismic waves are faster in cool areas of the mantle near subduction zones.

Locating an Earthquake's Epicenter

An instrument called a **seismometer** (size MAH muh ter) *measures and records ground motion and can be used to determine the distance seismic waves travel.* Ground motion is recorded as a **seismogram**, *a graphical illustration of seismic waves,* shown in **Figure 6.**

Seismologists use a method called triangulation to locate an earthquake's epicenter. This method uses the speeds and travel times of seismic waves to determine the distance to the earthquake epicenter from at least three different seismometers.

❶ Find the arrival time difference.

First, determine the number of seconds between the arrival of the first P-wave and the first S-wave on the seismogram. This time difference is called lag time. Using the time scale on the bottom of the seismogram, subtract the arrival time of the first P-wave from the arrival time of the first S-wave.

❷ Find the distance to the epicenter.

Next, use a graph showing the P-wave and S-wave lag time plotted against distance. Look at the *y*-axis and locate the place on the solid blue line that intersects with the lag time that you calculated from the seismogram. Then, read the corresponding distance from the epicenter on the *x*-axis.

❸ Plot the distance on a map.

Next, use a ruler and a map scale to measure the distance between the seismometer and the earthquake epicenter. Draw a circle with a radius equal to this distance by placing the compass point on the seismometer location. Set the pencil at the distance measured on the scale. Draw a complete circle around the seismometer location. The epicenter is somewhere on the circle. When circles are plotted for data from at least three seismic stations, the epicenter's location can be found. This location is the point where the three circles intersect.

Triangulation

❶ Find the arrival time difference.

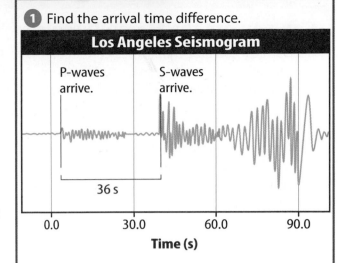

❷ Find the distance to the epicenter.

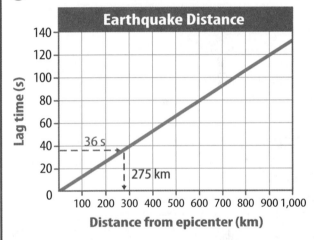

❸ Plot the distance on the map.

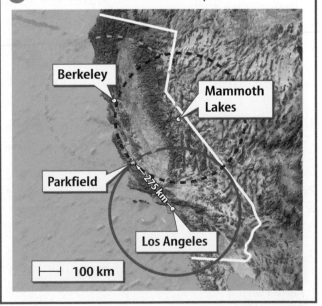

Figure 6 Seismograms provide the information necessary to locate an earthquake epicenter.

Inquiry MiniLab

15 minutes

Can you use the Mercalli scale to locate an epicenter?

Isoseismic (I soh SIZE mihk) lines connect areas that experience equal intensity during an earthquake. In this activity, you will observe trends in intensity and use the Mercalli scale to locate an earthquake epicenter.

1. Obtain a **map** of Mercalli ratings for the San Francisco Bay area.

2. Draw a line that connects all the points of equal intensity, making a closed loop. This is your first isoseismic line.

3. Continue drawing isoseismic lines for each Mercalli rating on the map. Just like contour lines, these lines should never cross.

Analyze and Conclude

1. **Interpret Data** Identify two cities that experienced similar effects during the earthquake.

2. **Infer** What were some of the experiences people in San Francisco might have had during the earthquake?

3. 🔑 **Key Concept** Can you identify the earthquake's epicenter on your map? Why did you choose this location?

Math Skills

Use Roman Numerals
Use the following rules to evaluate Roman numerals.

1. Values: X = 10; V = 5; I = 1

2. Add similar values that are next to one another, such as III (1 + 1 + 1 = 3)

3. Add a smaller value that comes after a larger value, such as XV (10 + 5 = 15)

4. Subtract a smaller value that precedes a larger value, such as IX (10 − 1 = 9)

5. Use the fewest possible numerals to express the value (X rather than VV)

Practice
What is the value of the Roman numeral XVI? XIV?

- **Math Practice**
- **Personal Tutor**

Determining Earthquake Magnitude

Scientists can use three different scales to measure and describe earthquakes. The Richter magnitude scale uses the amount of ground motion at a given distance from an earthquake to determine magnitude. The Richter magnitude scale is used when reporting earthquake activity to the general public.

The Richter scale begins at zero, but there is no upper limit to the scale. Each increase of 1 unit on the scale represents ten times the amount of ground motion recorded on a seismogram. For example, a magnitude 8 earthquake produces 10 times greater shaking than a magnitude 7 earthquake and 100 times greater shaking than a magnitude 6 earthquake does. The largest earthquake ever recorded was a magnitude 9.5 in Chile in 1960. The earthquake and the tsunamis that followed left nearly 2,000 people dead and 2 million people homeless.

Seismologists use the moment magnitude scale to measure the total amount of energy released by the earthquake. The energy released depends on the size of the fault that breaks, the motion that occurs along the fault, and the strength of the rocks that break during an earthquake. The units on this scale are exponential. For each increase of one unit on the scale, the earthquake releases 31.5 times more energy. That means that a magnitude 8 earthquake releases more than 992 times the amount of energy than that of a magnitude 6 earthquake.

Reading Check Compare the Richter scale to the moment magnitude scale.

Describing Earthquake Intensity

Another way to measure and describe an earthquake is to evaluate the damage that results from shaking. Shaking is directly related to earthquake intensity. The Modified Mercalli scale measures earthquake intensity based on descriptions of the earthquake's effects on people and structures. The Modified Mercalli scale, shown in **Table 3,** ranges from I, when shaking is not noticeable, to XII, when everything is destroyed.

Local geology also contributes to earthquake damage. In an area covered by loose sediment, ground motion is exaggerated. The intensity of the earthquake will be greater there than in places built on solid bedrock even if they are the same distance from the epicenter. Recall the lesson opener. The 1989 Loma Prieta earthquake produced severe shaking in an area called the Marina District in the San Francisco Bay area. This area had been built on loose sediment susceptible to shaking.

Table 3 The Modified Mercalli scale is used to evaluate earthquake intensity based on the damage that results.

Table 3	Modified Mercalli Scale
I	Not felt except under unusual conditions.
II	Felt by few people; suspended objects might swing.
III	Most noticeable indoors; vibrations feel like the effects of a truck passing by.
IV	Felt by many people indoors but by few people outdoors; dishes and windows rattle; standing cars rock noticeably.
V	Felt by nearly everyone; some dishes and windows break and some walls crack.
VI	Felt by all; furniture moves; some plaster falls from walls and some chimneys are damaged.
VII	Everybody runs outdoors; some chimneys break; damage is light in well-built structures but considerable in weak structures.
VIII	Chimneys, smokestacks, and walls fall; heavy furniture is overturned; partial collapse of ordinary buildings occurs.
IX	Great general damage occurs; buildings shift off foundations; ground cracks; underground pipes break.
X	Most ordinary structures are destroyed; rails are bent; landslides are common.
XI	Few structures remain standing; bridges are destroyed; railroad rails are greatly bent; broad fissures form in the ground.
XII	Total destruction; objects are thrown upward into the air.

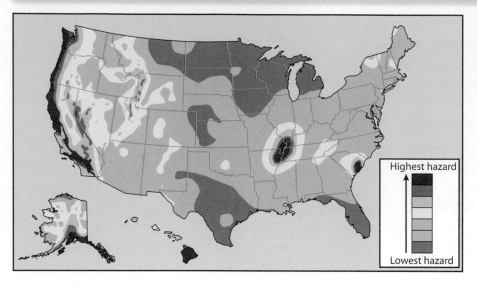

Figure 7 Areas that experienced earthquakes in the past will likely experience earthquakes again. Notice that even some parts of the central and eastern United States have high earthquake risk because of past activity.

Highest hazard

Lowest hazard

REVIEW VOCABULARY ·····

convergent
tending to move toward one point or approaching each other

Earthquake Risk

Recall that most earthquakes occur near tectonic plate boundaries. The transform plate boundary in California and the convergent plate boundaries in Oregon, Washington, and Alaska have the highest earthquake risks in the United States. However, not all earthquakes occur near plate boundaries. Some of the largest earthquakes in the United States have occurred far from plate boundaries.

From 1811–1812, three earthquakes with magnitudes between 7.8 and 8.1 occurred on the New Madrid Fault in Missouri. In contrast, the 1989 Loma Prieta earthquake had a magnitude of 7.1. **Figure 7** illustrates earthquake risk in the United States. Fortunately, high energy, destructive earthquakes are not very common. On average, only about 10 earthquakes with a magnitude greater than 7.0 occur worldwide each year. Earthquakes with magnitudes greater than 9.0, such as the Indian Ocean earthquake that caused the Asian tsunami in 2004, are rare.

Because earthquakes threaten people's lives and property, seismologists study the probability that an earthquake will occur in a given area. Probability is one of several factors that contribute to earthquake risk assessment. Seismologists also study past earthquake activity, the geology around a fault, the population density, and the building design in an area to evaluate risk. Engineers use these risk assessments to design earthquake-safe structures that are able to withstand the shaking during an earthquake. City and state governments use risk assessments to help plan and prepare for future earthquakes.

 Key Concept Check How do seismologists evaluate risk?

Lesson 1 Review

Visual Summary

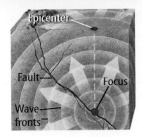

The focus is the area on a fault where an earthquake begins.

Earthquakes can occur along plate boundaries.

Seismologists assess earthquake risk by studying past earthquake activity and local geology.

FOLDABLES

Use your lesson Foldable to review the lesson. Save your Foldable for the project at the end of the chapter.

What do you think NOW?

You first read the statements below at the beginning of the chapter.

1. Earth's crust is broken into rigid slabs of rock that move, causing earthquakes and volcanic eruptions.

2. Earthquakes create energy waves that travel through Earth.

3. All earthquakes occur on plate boundaries.

Did you change your mind about whether you agree or disagree with the statements? Rewrite any false statements to make them true.

Use Vocabulary

1 **Compare and contrast** the three types of faults.

2 **Distinguish** between an earthquake focus and an earthquake epicenter.

3 **Use the terms** *seismogram* and *seismometer* in a sentence.

Understand Key Concepts

4 **Identify** areas in the United States that have the highest earthquake risk.

5 Approximately how much more energy is released in a magnitude 7 earthquake compared to a magnitude 5 earthquake?
  A. 30   C. 90
  B. 60   D. 1000

Interpret Graphics

6 **Compare and contrast** Create a table with the column headings for wave type, wave motion, and wave properties. Use the table to compare and contrast the three types of seismic waves.

7 **Describe** Use the image below to describe Earth's interior.

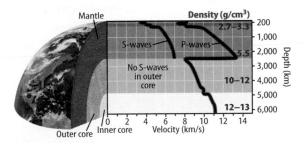

Critical Thinking

8 **Determine** what measurements you would make to evaluate earthquake risk in your hometown.

Math Skills   Review
— Math Practice —

9 What is the value of Roman numeral XXVI?

Can you locate an earthquake's epicenter?

Imagine the room where you are sitting suddenly begins to shake. This movement lasts for about 10 seconds. Based on the shaking that you have felt, it might seem like an earthquake happened nearby. But, to locate the epicenter, you need to analyze P-wave and S-wave data recorded for the same earthquake in at least three different locations.

Materials

map of North America

drawing compass

Safety

Learn It

When scientists conduct experiments, they make measurements and collect and **analyze** data. For example, seismologists measure the difference in arrival times between P-waves and S-waves following an earthquake. They collect seismic wave data from at least three different locations. Using the difference in arrival times, or lag time, seismologists can determine the distance to an earthquake epicenter.

Try It

1. Read and complete a lab safety form.

2. Obtain a map of the United States from your teacher.

3. Study the three seismograms. Determine the arrival times, to the nearest second, of the P- and S-waves for each seismometer station: Berkeley, CA; Parkfield, CA; and Kanab, UT. Record the location and the arrival times for P- and S-waves in your Science Journal.

4. Subtract the P-wave arrival time from the S-wave arrival time and record the lag time in your Science Journal.

5. Use the lag time and the Earthquake Distance graph to determine the distance to the epicenter for each seismometer station.

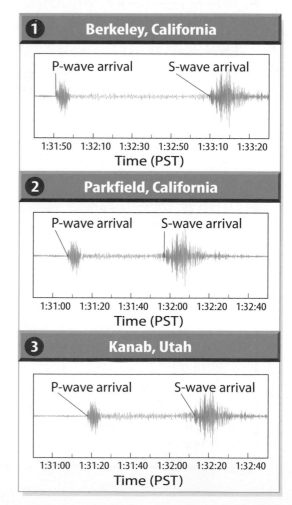

6 Use the map scale to set the spacing between the pencil and the point on the compass equal to the distance to the first seismometer. Draw a circle with a radius equal to the distance around the seismic station on the map.

7 Repeat for the two other seismometer locations. The point where the three circles intersect marks the earthquake epicenter.

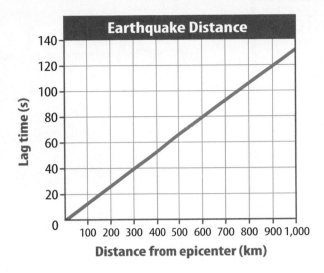

Earthquake Distance

Lag time (s) vs. Distance from epicenter (km)

Apply It

8 Consider the difference between the arrival times of P-waves for all three seismometer locations. Why does this difference occur?

9 **Examine** the calculated lag times for all three seismograms. Why do you think the arrival-time differences are greater for the stations that are furthest from the epicenter?

10 Where did the earthquake occur?

11 **Key Concept** Why does it take three seismograms to locate an earthquake epicenter? What is this process called?

Volcanoes

Reading Guide

Key Concepts 🔑
ESSENTIAL QUESTIONS

- How do volcanoes form?
- What factors contribute to the eruption style of a volcano?
- How are volcanoes classified?

Vocabulary

volcano p. 307

magma p. 307

lava p. 308

hot spot p. 308

shield volcano p. 310

composite volcano p. 310

cinder cone p. 310

volcanic ash p. 311

viscosity p. 311

  Multilingual eGlossary

🎬 **Video**

- **Science Video**
- **What's Science Got to do With It?**

Inquiry What makes an eruption explosive?

Notice the red, hot "fire fountain" erupting from Kilauea volcano in Hawaii. Kilauea is the most active volcano in the world. Now recall the ash eruption pictured in the chapter opener. What makes volcanoes erupt so differently? The answer can be found in magma chemistry.

What determines the shape of a volcano?

Not all volcanoes look the same. The location of a volcano and the magma chemistry play an important part in determining the shape of a volcano.

1. Read and complete a lab safety form.
2. Obtain a **tray, a beaker of sand, a beaker with a mixture of flour and water, waxed paper,** and a **plastic spoon.**
3. Lay the waxed paper inside the tray.
4. Hold the beaker of sand about 30 cm above the tray. Slowly pour the sand onto the waxed paper and observe how it piles up.
5. Fold the paper in half and use it to carefully pour the sand back into the beaker.
6. Stir the flour and water mixture. It should be about the consistency of oatmeal. Add water if necessary.
7. Repeat steps 4 and 5 with the flour and water mixture. Record your observations for each trial in your Science Journal.

Think About This

1. What do the sand and the flour and water mixture represent?

2. 🔑 **Key Concept** How do you think volcanoes get their shape?

What is a volcano?

Perhaps you have heard of some famous volcanoes such as Mount St. Helens, Kilauea, or Mount Pinatubo. All of these volcanoes have erupted within the last 30 years. A **volcano** *is a vent in Earth's crust through which melted—or molten—rock flows. Molten rock below Earth's surface is called* **magma.** Volcanoes are in many places worldwide. Some places have more volcanoes than others. In this lesson, you will learn about how volcanoes form, where they form, and about their structure and eruption style.

✓ **Reading Check** What is magma?

How do volcanoes form?

Volcanic eruptions constantly shape Earth's surface. They can form large mountains, create new crust, and leave a path of destruction behind. Scientists have learned that the movement of Earth's tectonic plates causes the formation of volcanoes and the eruptions that result.

◀ **Figure 8**
During subduction, magma forms when one plate sinks beneath another plate.

Figure 9 When plates spread apart, it forces magma to the surface and creates new crust. The pillow lava shown in the photograph formed at the mid-ocean ridge. ▼

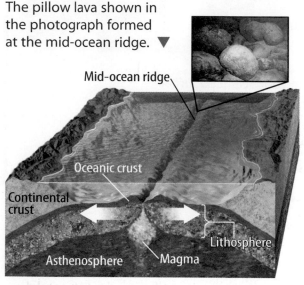

◀ **Figure 10**
The farther each of the Hawaiian Islands is from the hot spot, the older the island is.

Convergent Boundaries

Volcanoes can form along convergent plate boundaries. Recall that when two plates collide, the denser plate sinks, or subducts, into the mantle, as shown in **Figure 8.** The thermal energy below the surface and fluids driven off the subducting plate melt the mantle and form magma. Magma is less dense than the surrounding mantle and rises through cracks in the crust. This forms a volcano. *Molten rock that erupts onto Earth's surface is called* **lava.**

Divergent Boundaries

Lava erupts along divergent plate boundaries too. Recall that two plates spread apart along a divergent plate boundary. As the plates separate, magma rises through the vent or opening in Earth's crust that forms between them. This process commonly occurs at a mid-ocean ridge and forms new oceanic crust, as shown in **Figure 9.** More than 60 percent of all volcanic activity on Earth occurs along mid-ocean ridges.

Hot spots

Not all volcanoes form on or near plate boundaries. Volcanoes in the Hawaiian Island-Emperor Seamount chain are far from plate boundaries. *Volcanoes that are not associated with plate boundaries are called* **hot spots.** Geologists hypothesize that hot spots originate above a rising convection current from deep within Earth's mantle. They use the word *plume* to describe these rising currents of hot mantle material.

Figure 10 illustrates how a new volcano forms as a tectonic plate moves over a plume. When the plate moves away from the plume, the volcano becomes dormant, or inactive. Over time, a chain of volcanoes forms as the plate moves. The oldest volcano will be farthest away from the hot spot. The youngest volcano will be directly above the hot spot.

 Key Concept Check How do volcanoes form?

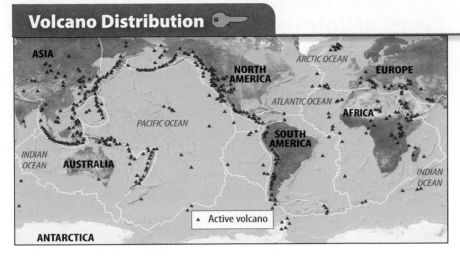

ASIA
ARCTIC OCEAN
NORTH AMERICA
EUROPE
ATLANTIC OCEAN
AFRICA
PACIFIC OCEAN
SOUTH AMERICA
INDIAN OCEAN
AUSTRALIA
INDIAN OCEAN
▲ Active volcano
ANTARCTICA

◀ **Figure 11** Most of the world's active volcanoes are located along convergent and divergent plate boundaries and hot spots.

Where do volcanoes form?

The world's active volcanoes are shown in **Figure 11.** The volcanoes all erupted within the last 100,000 years. Notice that most volcanoes are close to plate boundaries.

Ring of Fire

The Ring of Fire represents an area of earthquake and volcanic activity that surrounds the Pacific Ocean. When you compare the locations of active volcanoes and plate boundaries in **Figure 11,** you can see that volcanoes are mostly along convergent plate boundaries where plates collide. They also are located along divergent plate boundaries where plates separate. Volcanoes also can occur over hot spots, like Hawaii, the Galapagos Islands, and Yellowstone National Park in Wyoming.

 Reading Check Where is the Ring of Fire?

Volcanoes in the United States

There are 60 potentially active volcanoes in the United States. Most of these volcanoes are part of the Ring of Fire. Alaska, Hawaii, Washington, Oregon, and northern California all have active volcanoes, such as Mount Redoubt in Alaska. A few of these volcanoes have produced violent eruptions, like the explosive eruption of Mount St. Helens in 1980.

The United States Geological Survey (USGS) has established three volcano observatories to monitor the potential for future volcanic eruptions in the United States. Because large populations of people live near volcanoes such as Mount Rainier in Washington, shown in **Figure 12,** the USGS has developed a hazard assessment program. Scientists monitor earthquake activity, changes in the shape of the volcano, gas emissions, and the past eruptive history of a volcano to evaluate the possibility of future eruptions.

Figure 12 Mount Rainier is an active volcano in the Cascade Mountains of the Pacific Northwest. Many people live in close proximity to the volcano. ▼

Types of Volcanoes

Volcanoes are classified based on their shapes and sizes, as shown in **Table 4.** Magma composition and eruptive style of the volcano contribute to the shape. **Shield volcanoes** *are common along divergent plate boundaries and oceanic hot spots. Shield volcanoes are large with gentle slopes of basaltic lavas.* **Composite volcanoes** *are large, steep-sided volcanoes that result from explosive eruptions of andesitic and rhyolitic lava and ash along convergent plate boundaries.* **Cinder cones** *are small, steep-sided volcanoes that erupt gas-rich, basaltic lavas.* Some volcanoes are classified as supervolcanoes—volcanoes that have very large and explosive eruptions. Approximately 630,000 years ago, the Yellowstone Caldera in Wyoming ejected more than 1,000 km³ of rhyolitic ash and rock in one eruption. This eruption produced nearly 2,500 times the volume of material erupted from Mount St. Helens in 1980.

Table 4 Geologists classify volcanoes based on their size, shape, and eruptive style.

 Key Concept Check What determines the shape of a volcano?

Table 4 Volcanic Features

Concepts in Motion Interactive Table

Shield volcano

Large, shield-shaped volcano with gentle slopes made from basaltic lavas.

Composite volcano

Large, steep-sided volcano made from a mixture of andesitic and rhyolitic lava and ash.

Cinder cone volcano

Small, steep-sided volcano; made from moderately explosive eruptions of basaltic lavas.

Caldera

Large volcanic depression formed when a volcano's summit collapses or is blown away by explosive activity.

Volcanic Eruptions

When magma surfaces, it might erupt as a lava flow, such as the lava shown in **Figure 13** erupting from Kilauea volcano in Hawaii. Other times, magma might erupt explosively, sending **volcanic ash**—*tiny particles of pulverized volcanic rock and glass*—high into the atmosphere. **Figure 13** also shows Mount St. Helens in Washington, erupting violently in 1980. Why do some volcanoes erupt violently while others erupt quietly?

Eruption Style

Magma chemistry determines a volcano's eruptive style. The explosive behavior of a volcano is affected by the amount of dissolved gases, specifically the amount of water vapor, a magma contains. It is also affected by the silica, SiO_2, content of magma.

Magma Chemistry Magmas that form in different volcanic environments have unique chemical compositions. Silica is the main chemical compound in all magmas. Differences in the amount of silica affect magma thickness and its **viscosity**—*a liquid's resistance to flow.*

Magma that has a low silica content also has a low viscosity and flows easily like warm maple syrup. When the magma erupts, it flows as fluid lava that cools, crystallizes, and forms the volcanic rock basalt. This type of lava commonly erupts along mid-ocean ridges and at oceanic hot spots, such as Hawaii.

Magma that has a high silica content has a high viscosity and flows like sticky toothpaste. This type of magma forms when rocks rich in silica melt or when magma from the mantle mixes with continental crust. The volcanic rocks andesite and rhyolite form when intermediate and high silica magmas erupt from subduction zone volcanoes and continental hot spots.

 Key Concept Check What factors affect eruption style?

Quiet Eruption

Violent Eruption

Figure 13 Lavas that are low in silica and the amount of dissolved gases erupt quietly. Explosive eruptions result from lava and ash that are high in silica and dissolved gases.

Dissolved Gases

Dissolved Gases The presence of **dissolved** gases in magma contributes to how explosive a volcano can be. This is similar to what happens when you shake a can of soda and then open it. The bubbles come from the carbon dioxide that is dissolved in the soda. The pressure inside the can decreases rapidly when you open it. Trapped bubbles increase in size rapidly and escape as the soda erupts from the can.

All magmas contain dissolved gases. These gases include water vapor and small amounts of carbon dioxide and sulfur dioxide. As magma moves toward the surface, the pressure from the weight of the rock above decreases. As pressure decreases, the ability of gases to stay dissolved in the magma also decreases. Eventually, gases can no longer remain dissolved in the magma and bubbles begin to form. As the magma continues to rise to the surface, the bubbles increase in size and the gas begins to escape. Because gases cannot easily escape from high-viscosity lavas, this combination often results in explosive eruptions. When gases escape above ground, the lava, ash, or volcanic glass that cools and crystallizes has holes. These holes, shown in **Figure 14**, are a common feature in the volcanic rock pumice.

ACADEMIC VOCABULARY
dissolve
(verb) to cause to disperse or disappear

Figure 14 The holes in this pumice were caused by gas bubbles that escaped during a volcanic eruption.

Inquiry) MiniLab   **20 minutes**

Can you model the movement of magma?

Magma erupts because it is less dense than Earth's crust. Similarly, oil is less dense than water and can be used to model magma.

1. Read and complete a lab safety form.
2. Half-fill a **clear plastic cup** with **pebbles.**
3. Fill the cup with **water** to a level just above the top of the pebbles.
4. Fill a **syringe** with 5 mL of **olive oil.**

5. Insert the syringe between the pebbles and the side of the cup until it touches the bottom.
6. Inject the oil slowly, 1 mL at a time.
7. Observe and record your results in your Science Journal.
8. Repeat the procedures using **motor oil.**

Analyze and Conclude

1. **Observe** What happens to the oil when you inject it into the water?

2. **Compare** How did the movement of the two oils differ?

3. 🔑 **Key Concept** Which oil behaves like magma that will become basalt? Which behaves like magma that will become rhyolite? Explain.

Effects of Volcanic Eruptions

On average, about 60 different volcanoes erupt each year. The effects of lava flows, ash fall, pyroclastic flows, and mudflows can affect all life on Earth. Volcanoes enrich rock and soil with valuable nutrients and help to regulate climate. Unfortunately, they also can be destructive and sometimes even deadly.

Lava Flows Because lava flows are relatively slow moving, they are rarely deadly. But lava flows can be damaging. Mount Etna in Sicily, Italy, is Europe's most active volcano. **Figure 15** shows a fountain of fluid, hot lava erupting from one of the volcano's many vents. In May 2008, the volcano began spewing lava and ash in an eruption lasting over six months. Although lavas tend to be slow moving, they threaten communities nearby. People who live on Mount Etna's slopes are used to evacuations due to frequent eruptions.

Ash Fall During an explosive eruption, volcanoes can erupt large volumes of volcanic ash. Ash columns can reach heights of more than 40 km. Recall that ash is a mixture of particles of pulverized rock and glass. Ash can disrupt air traffic and cause engines to stop mid-flight as shards of rock and ash fuse onto hot engine blades. Ash can also affect air quality and can cause serious breathing problems. Large quantities of ash erupted into the atmosphere can also affect climate by blocking out sunlight and cooling Earth's atmosphere.

Mudflows The thermal energy a volcano produces during an eruption can melt snow and ice on the summit. This meltwater can then mix with mud and ash on the mountain to form mudflows. Mudflows are also called lahars. Mount Redoubt in Alaska erupted on March 23, 2009. Snow and meltwater mixed to form the mudflows shown in **Figure 16**.

▲ **Figure 15** Mount Etna is one of the world's most active volcanoes. People that live near the volcano are accustomed to frequent eruptions of both lava and ash.

◀ **Figure 16** Many of the steep-sided composite volcanoes are covered with seasonal snow. When a volcano becomes active, the snow can melt and mix with mud and ash to form a mudflow like the one shown here in the Cook Inlet, Alaska.

▲ **Figure 17** A pyroclastic flow travels down the side of Mount Mayon in the Philippines. Pyroclastic flows are made of hot (*pyro*) volcanic particles (*clast*).

Pyroclastic Flow Explosive volcanoes can produce fast-moving avalanches of hot gas, ash, and rock called pyroclastic (pi roh KLAS tihk) flows. Pyroclastic flows travel at speeds of more than 100 km/hr and with temperatures greater than 1000°C. In 1980, Mount St. Helens produced a pyroclastic flow that killed 58 people and destroyed 1 billion km³ of forest. Mount Mayon in the Phillippines erupts frequently producing pyroclastic flows like the one shown in **Figure 17.**

Predicting Volcanic Eruptions

Unlike earthquakes, volcanic eruptions can be predicted. Moving magma can cause ground deformation, a change in shape of the volcano, and a series of earthquakes called an earthquake swarm. Volcanic gas emissions can increase. Ground and surface water near the volcano can become more acidic. Geologists study these events, in addition to satellite and aerial photographs, to assess volcanic hazards.

Volcanic Eruptions and Climate Change

Volcanic eruptions affect climate when volcanic ash in the atmosphere blocks sunlight. High-altitude wind can move ash around the world. In addition, sulfur dioxide gases released from a volcano form sulfuric acid droplets in the upper atmosphere. These droplets reflect sunlight into space, resulting in lower temperatures as less sunlight reaches Earth's surface. **Figure 18** shows the result of sulfur dioxide gas in the atmosphere from the 1991 eruption of Mt. Pinatubo.

Key Concept Check How do volcanoes affect climate?

Figure 18 In 1991, Mount Pinatubo erupted more than 20 million tons of gas and volcanic ash into the atmosphere. The greatest concentration of sulfur dioxide gas from the eruption is shown below in blue. The eruption caused temperatures to decrease by almost one degree Celsius in one year. ▼

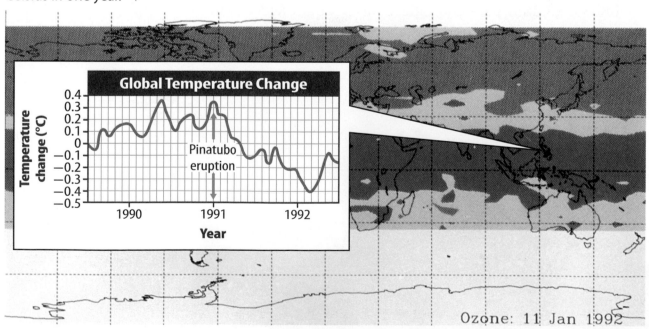

Ozone: 11 Jan 1992

Lesson 2 Review

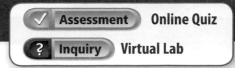
Visual Summary

Volcanoes form when magma rises through cracks in the crust and erupts from vents on Earth's surface.

Magma with low amounts of silica and low viscosity erupts to form shield volcanoes.

Magma with high amounts of silica and high viscosity erupts explosively to form composite cones.

FOLDABLES®

Use your lesson Foldable to review the lesson. Save your Foldable for the project at the end of the chapter.

What do you think NOW?

You first read the statements below at the beginning of the chapter.

4. Volcanoes can erupt anywhere on Earth.

5. Volcanic eruptions are rare.

6. Volcanic eruptions only affect people and places close to the volcano.

Did you change your mind about whether you agree or disagree with the statements? Rewrite any false statements to make them true.

Use Vocabulary

1 **Compare and contrast** lava and magma.

2 **Explain** the term *viscosity*.

3 Pulverized rock and ash that erupts from explosive volcanoes is called _____.

Understand Key Concepts

4 **Identify** places where volcanoes form.

5 **Compare** the three main types of volcanoes.

6 What type of lava erupts from shield volcanoes?
   A. andesitic   C. granitic
   B. basaltic   D. rhyolitic

Interpret Graphics

7 **Analyze** the image below and explain what factors contribute to explosive eruptions.

8 **Create** a graphic organizer to illustrate the four types of eruptive products that can result from a volcanic eruption.

Eruptive Products

Critical Thinking

9 **Compare** the shapes of composite volcanoes and shield volcanoes. Why are their shapes and eruptive styles so different?

10 **Explain** how explosive volcanic eruptions can cause climate change. What might happen if Yellowstone Caldera erupted today?

Record your answers on the answer sheet provided by your teacher or on a sheet of paper.

Multiple Choice

1 Along which type of plate boundary do the deepest earthquakes occur?

　A convergent

　B divergent

　C passive

　D transform

2 The Richter scale registers the magnitude of an earthquake by determining the

　A amount of energy released by the earthquake.

　B amount of ground motion measured at a given distance from the earthquake.

　C descriptions of damage caused by the earthquake.

　D type of seismic waves produced by the earthquake.

3 Which state has no active volcanoes?

　A California

　B Hawaii

　C New York

　D Washington

Use the diagram below to answer question 4.

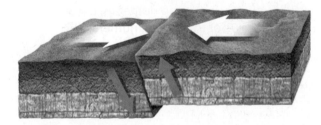

4 Which type of fault is shown in the diagram above?

　A normal

　B reverse

　C shallow

　D strike-slip

Use the diagram below to answer question 5.

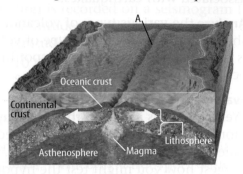

5 Which feature is labeled with the letter *A* in the diagram above?

　A a caldera

　B a chain of hot spot volcanoes

　C a mid-ocean ridge

　D a subducting tectonic plate

6 Which term describes a fast-moving avalanche of hot gas, ash, and rock that erupts from an explosive volcano?

　A ash fall

　B cinder cone

　C lahar

　D pyroclastic flow

7 Earthquakes occur along the San Andreas Fault. Which is an example of this type of plate boundary?

　A convergent

　B divergent

　C passive

　D transform

8 Hot spot volcanoes ALWAYS

　A appear at plate boundaries.

　B erupt in chains.

　C form above mantle plumes.

　D remain active.

Use the map below to answer questions 9 and 10.

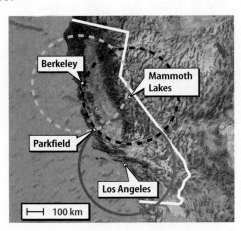

9 What do the circles represent in the map of seismic activity illustrated above?

A the distance between waves

B the distance to an earthquake epicenter

C the seismic wave speeds

D the wave travel times

10 According to the map, where is the earthquake epicenter?

A Berkeley

B Los Angeles

C Mammoth Lakes

D Parkfield

11 Where do seismic waves originate?

A above ground

B epicenter

C focus

D seismogram

Constructed Response

Use the diagram below to answer questions 12 and 13.

12 The diagram above shows one way volcanoes form. Explain the process shown in the diagram and why volcanoes form as a result of this process.

13 What type of volcano results from the process shown in the diagram? Describe it. What is the eruptive style of this type of volcano? Why?

Use the table below to answer question 14.

Wave Type	Characteristics

14 Re-create the table above and identify the three types of seismic waves. Then, describe wave characteristics such as movement, speed, and difference in arrival time for each type.

NEED EXTRA HELP?													
If You Missed Question...	1	2	3	4	5	6	7	8	9	10	11	12	13	14
Go to Lesson...	1	1	2	1	2	2	1	2	1	1	1	2	2	1

Chapter 10

Clues to Earth's Past

THE BIG IDEA

What evidence do scientists use to determine the ages of rocks?

Inquiry Always a Canyon?

The Colorado River started cutting through the rock layers of the Grand Canyon only about 6 million years ago. Hundreds of millions of years earlier, these rock layers were deposited at the bottom of an ancient sea. Even before that, a huge mountain range existed here.

- What evidence do scientists use to learn about past environments?

- What evidence do scientists use to determine the ages of rocks?

Get Ready to Read

What do you think?

Before you read, decide if you agree or disagree with each of these statements. As you read this chapter, see if you change your mind about any of the statements.

1 Fossils are pieces of dead organisms.

2 Only bones can become fossils.

3 Older rocks are always located below younger rocks.

4 Relative age means that scientists are relatively sure of the age.

5 Absolute age means that scientists are sure of the age.

6 Scientists use radioactive decay to determine the ages of some rocks.

ConnectED Your one-stop online resource

connectED.mcgraw-hill.com

Video

WebQuest

Audio

Assessment

Review

Concepts in Motion

Inquiry

Multilingual eGlossary

Fossils

Reading Guide

Key Concepts 🔑
ESSENTIAL QUESTIONS

- What are fossils and how do they form?
- What can fossils reveal about Earth's past?

Vocabulary

fossil p. 327

catastrophism p. 327

uniformitarianism p. 328

carbon film p. 330

mold p. 331

cast p. 331

trace fossil p. 331

paleontologist p. 332

g Multilingual eGlossary

▢ Video

- BrainPOP®
- Science Video

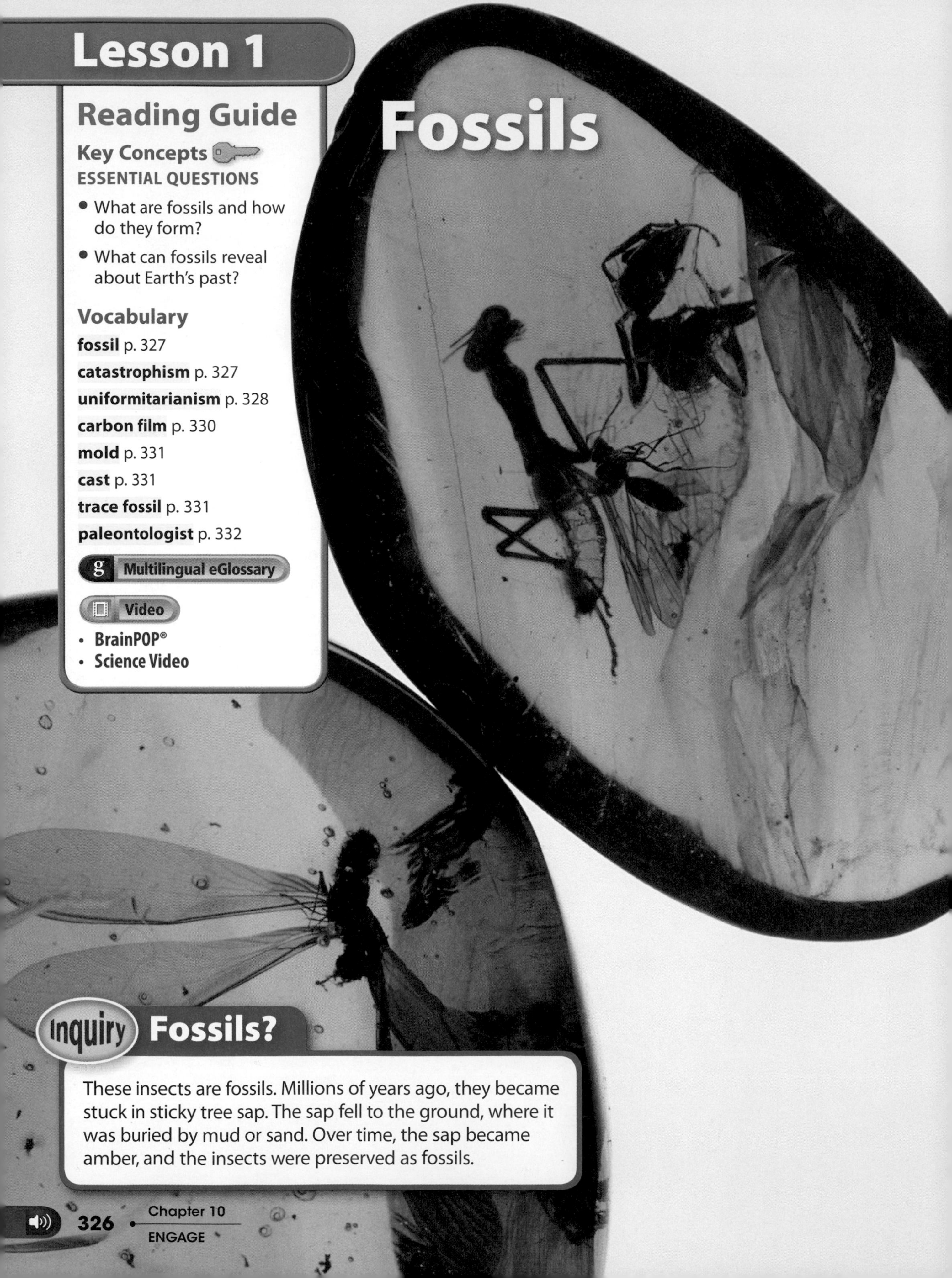

Inquiry Fossils?

These insects are fossils. Millions of years ago, they became stuck in sticky tree sap. The sap fell to the ground, where it was buried by mud or sand. Over time, the sap became amber, and the insects were preserved as fossils.

What can trace fossils show?

Did you know that a fossil can be a footprint or the imprint of an ancient nest? These are examples of trace fossils. Although trace fossils do not contain any part of an organism, they do hold clues about how organisms lived, moved, or behaved.

1 Read and complete a lab safety form.

2 Flatten some **clay** into a pancake shape.

3 Think about a behavior or movement you would like your fossil to model. Use available tools, such as a **plastic knife,** a **chenille stem,** or a **toothpick,** to make a fossil showing that behavior or movement.

4 Exchange your fossil with another student. Try to figure out what behavior or movement he or she modeled.

Think About This

1. Were you able to determine what behavior or movement your classmate's fossil modeled? Was he or she able to determine yours? Why or why not?

2. 🔑 **Key Concept** What do you think scientists can learn by studying trace fossils?

Evidence of the Distant Past

Have you ever looked through an old family photo album? Each photo shows a little of your family's history. You might guess the age of the photographs based on the clothes people are wearing, the vehicles they are driving, or even the paper the photographs are printed on.

Just as old photos can provide clues to your family's past, rocks can provide clues to Earth's past. Some of the most obvious clues found in rocks are the remains or traces of ancient living things. **Fossils** *are the preserved remains or evidence of ancient living things.*

Catastrophism

Many fossils represent plants and animals that no longer live on Earth. Ideas about how these fossils formed have changed over time. Some early scientists thought that great, sudden, catastrophic disasters killed the organisms that became fossils. These scientists explained Earth's history as a series of disastrous events occurring over short periods of time. **Catastrophism** (kuh TAS truh fih zum) *is the idea that conditions and organisms on Earth change in quick, violent events.* The events described in catastrophism include volcanic eruptions and widespread flooding. Scientists eventually disagreed with catastrophism because Earth's history is full of violent events.

WORD ORIGIN ············

fossil
from Latin *fossilis,* means "dug up"

Figure 1 Hutton realized that erosion happens on small or large scales.

ACADEMIC VOCABULARY

uniform
(adjective) having always the same form, manner, or degree; not varying or variable

Uniformitarianism

Most people who supported catastrophism thought that Earth was only a few thousand years old. In the 1700s, James Hutton rejected this idea. Hutton was a naturalist and a farmer in Scotland. He observed how the landscape on his farm gradually changed over the years. Hutton thought that the processes responsible for changing the landscape on his farm could also shape Earth's surface. For example, he thought that erosion caused by streams, such as that shown in **Figure 1,** could also wear down mountains. Because he realized that this would take a long time, Hutton proposed that Earth is much older than a few thousand years.

Hutton's ideas eventually were included in a principle called uniformitarianism (yew nuh for muh TER ee uh nih zum). *The principle of* **uniformitarianism** *states that geologic processes that occur today are similar to those that have occurred in the past.* According to this view, Earth's surface is constantly being reshaped in a steady, *uniform* manner.

✓ Reading Check What is uniformitarianism?

Today, uniformitarianism is the basis for understanding Earth's past. But scientists also know that catastrophic events do sometimes occur. Huge volcanic eruptions and giant meteorite impacts can change Earth's surface very quickly. These catastrophic events can be explained by natural processes.

Inquiry MiniLab **15 minutes**

How is a fossil a clue?

Fossils provide clues about once-living organisms. Sometimes those clues are hard to interpret.

1. Read and complete a lab safety form.
2. Select an **object** from a bag provided by your teacher. Do not let anyone see your object.
3. Make a fossil impression of your object by pressing only part of it into a piece of **clay.**
4. Place your clay fossil and object in separate locations indicated by your teacher.
5. Make a chart in your Science Journal that matches your classmates' objects and fossils.

Analyze and Conclude

1. Did you correctly match the objects with their fossils?

2. Why might scientists need more than one fossil of an organism to understand what it looked like?

3. 🔑 **Key Concept** What do you think you could learn from fossils?

Figure 2 A fossil can form if an organism with hard parts, such as a fish, is buried quickly after it dies.

1 A dead fish falls to a river bottom during a flood. Its body is rapidly buried by mud, sand, or other sediment.

2 Over time, the body decomposes but the hard bones become a fossil.

3 The sediments, hardened into rock, are uplifted and eroded, which exposes the fossil fish on the surface.

Formation of Fossils

Recall that fossils are the remains or traces of ancient living organisms. Not all dead organisms become fossils. Fossils form only under certain conditions.

Conditions for Fossil Formation

Most plants and animals are eaten or decay when they die, leaving no evidence that they ever lived. Think about the chances of an apple becoming a fossil. If it is on the ground for many months, it will decay into a soft, rotting lump. Eventually, insects and bacteria consume it.

However, some conditions increase the chances of fossil formation. An organism is more likely to become a fossil if it has hard parts, such as shells, teeth, or bones, like the fish in **Figure 2.** Unlike a soft apple, hard parts do not decay easily. Also, an organism is more likely to form a fossil if it is buried quickly after it dies. If layers of sand or mud bury an organism quickly, decay is slowed or stopped.

🔑 **Key Concept Check** What conditions increase the chances of fossil formation?

Fossils Come in All Sizes

You might have seen pictures of dinosaur fossils. Many dinosaurs were large animals, and large bones were left behind when they died. Not all fossils are large enough for you to see. Sometimes it is necessary to use a microscope to see fossils. Tiny fossils are called microfossils. The microfossils in **Figure 3** are each about the size of a speck of dust.

☑ **Visual Check** How did the fish fossil reach the surface?

Figure 3 Details of microfossils can be seen only under a microscope.

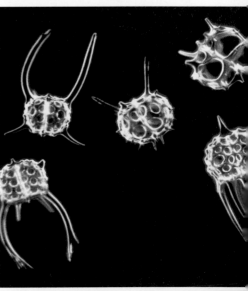

Types of Preservation

Fossils are preserved in different ways. As shown in **Figure 4,** there are many ways fossils can form.

Preserved Remains

Sometimes the actual remains of organisms are preserved as fossils. For this to happen, an organism must be completely enclosed in some material over a long period of time. This would prevent it from being exposed to air or bacteria. Generally, preserved remains are 10,000 or fewer years in age. However, insects preserved in amber—shown in the photo at the beginning of this lesson—can be millions of years old.

Carbon Films

Sometimes when an organism is buried, exposure to heat and pressure forces gases and liquids out of the organism's tissues. This leaves only the carbon behind. *A* **carbon film** *is the fossilized carbon outline of an organism or part of an organism.*

Mineral Replacement

Replicas, or copies, of organisms can form from minerals in groundwater. They fill in the pore spaces or replace the tissues of dead organisms. **Petrified** wood is an example.

Figure 4 Fossils can form in many different ways.

Types of Preservation 🔑

Preserved Remains Organisms trapped in amber, tar pits, or ice can be preserved over thousands of years. This baby mammoth was preserved in ice for more than 10,000 years before it was discovered. ▶

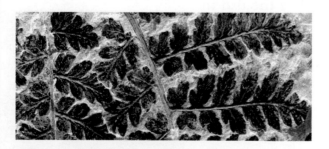

◀ **Carbon Film** Only a carbon film remains of this ancient fern. Carbon films are usually shiny black or brown. Fish, insects, and plant leaves are often preserved as carbon films.

Mineral Replacement Rock-forming minerals dissolved in groundwater can fill in pore spaces or replace the tissues of dead organisms. This petrified wood formed when silica (SiO_2) filled in the spaces between the cell walls in a dead tree. The wood petrified when the SiO_2 crystallized. ▶

Molds

Sometimes all that remains of an organism is its fossilized imprint or impression. *A **mold** is the impression in a rock left by an ancient organism.* A mold can form when sediment hardens around a buried organism. As the organism decays over time, an impression of its shape remains in the sediment. The sediment eventually turns to rock.

Casts

Sometimes, after a mold forms, it is filled with more sediment. *A **cast** is a fossil copy of an organism made when a mold of the organism is filled with sediment or mineral deposits.* The process is similar to making a gelatin dessert using a molded pan.

Trace Fossils

Some animals leave fossilized traces of their movement or activity. *A **trace fossil** is the preserved evidence of the activity of an organism.* Trace fossils include tracks, footprints, and nests. These fossils help scientists learn about characteristics and behaviors of animals. The dinosaur tracks in **Figure 4** reveal clues about the dinosaur's size, its speed, and whether it was traveling alone or in a group.

 Reading Check What are some examples of trace fossils?

FOLDABLES®

Make a tri-fold book from a sheet of paper. Label it as shown. Use it to organize your notes about the different types of fossils.

Types of fossils | How they formed | Examples

Mold This mold of an ancient mollusk formed after it was buried by sediment and then decayed. The sediment hardened, leaving an impression of its shape in the rock. ▼

▲ **Cast** This cast was formed when the mold was later filled with sediment that then hardened. Molds and casts show only the exterior, or outside, features of organisms.

Trace Fossil These trace fossils formed when dinosaur tracks in soft sediments were later filled in by other sediments, which then hardened. Trace fossils reveal information about the behavior of organisms. ▶

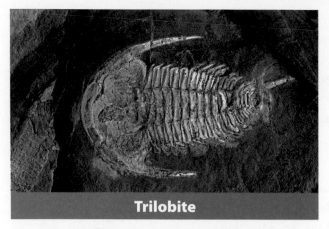

Trilobite

Horseshoe Crab

▲ **Figure 5** Partly because a trilobite fossil looks like a present-day horseshoe crab, scientists infer that the trilobite lived in an environment similar to the environment where a horseshoe crab lives.

Ancient Environments

Scientists who study fossils are called **paleontologists** (pay lee ahn TAH luh jihstz). Paleontologists use the principle of uniformitarianism to learn about ancient organisms and the environments where ancient organisms lived. For example, they can compare fossils of ancient organisms with organisms living today. The trilobite fossil and the horseshoe crab in **Figure 5** look alike. Horseshoe crabs today live in shallow water on the ocean floor. Partly because trilobite fossils look like horseshoe crabs, paleontologists infer that trilobites also lived in shallow ocean water.

Shallow Seas

Today, Earth's continents are mostly above sea level. But sea level has risen, flooding Earth's continents, many times in the past. For example, a shallow ocean covered much of North America 450 million years ago, as illustrated in the map in **Figure 6.** Fossils of organisms that lived in that shallow ocean, like those shown in **Figure 6,** help scientists reconstruct what the seafloor looked like at that time.

Key Concept Check What can fossils tell us about ancient environments?

Figure 6 Studying fossils helped scientists reconstruct this 450-million-year-old North American seafloor. Most of what would become the United States was covered by a shallow sea during that time.

Land
Shallow ocean
Deeper ocean

Figure 7 About 100 million years ago, tropical forests and swamps covered much of North America. Dinosaurs also lived on Earth at that time.

Past Climates

You might have heard people talking about global climate change, or maybe you've read about climate change. Evidence indicates that Earth's present-day climate is warming. Fossils show that Earth's climate has warmed and cooled many times in the past.

Plant fossils are especially good indicators of climate change. For example, fossils of ferns and other tropical plants dating to the time of the dinosaurs reveal that Earth was very warm 100 million years ago. Tropical forests and swamps covered much of the land, as illustrated in **Figure 7.**

 Key Concept Check What was Earth's climate like when dinosaurs lived?

Millions of years later, the swamps and forests were gone, but coarse grasses grew in their place. Huge sheets of ice called glaciers spread over parts of North America, Europe, and Asia. Fossils suggest that some species that lived during this time, such as the woolly mammoth shown in **Figure 8,** were able to survive in the colder climate.

Fossils of organisms such as ferns and mammoths help scientists learn about ancient organisms and past environments. In the following lessons, you will read how scientists use fossils and other clues, such as the order of rock layers and radioactivity, to learn about the ages of Earth's rocks.

The mammoth's huge teeth could grind the coarse grasses that grew in the cold climate.

Figure 8 The woolly mammoth was well adapted to a cold climate.

Lesson 1 Review

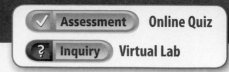
Visual Summary

The principle of uniformitarianism is the basis for understanding Earth's past.

Fossils can form in many different ways.

Fossils help scientists learn about Earth's ancient organisms and past environments.

FOLDABLES®

Use your lesson Foldable to review the lesson. Save your Foldable for the project at the end of the chapter.

What do you think NOW?

You first read the statements below at the beginning of the chapter.

1. Fossils are pieces of dead organisms.

2. Only bones can become fossils.

Did you change your mind about whether you agree or disagree with the statements? Rewrite any false statements to make them true.

Use Vocabulary

1. **Distinguish** between catastrophism and uniformitarianism.

2. Plant leaves are often preserved as _____.

3. **Use the terms** *cast* and *mold* in a complete sentence.

Understand Key Concepts 🗝

4. Which conditions aid in the formation of fossils?
   A. hard parts and slow burial
   B. hard parts and rapid burial
   C. soft parts and rapid burial
   D. soft parts and slow burial

5. What human body system could be fossilized? Explain.

6. **Determine** what type of an environment a fossil palm tree would indicate.

Interpret Graphics

7. **Compare** the two sets of dinosaur footprints below. Which dinosaur was running? How can you tell?

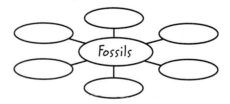

8. **Organize Information** Copy and fill in the graphic organizer below to list types of fossil preservation.

Fossils

Critical Thinking

9. **Invent** a process for the formation of ocean basins consistent with catastrophism.

10. **Evaluate** how the following statement relates to what you have read in this lesson: "The present is the key to the past."

▼ This Byronosaurus skull was discovered at Ukhaa Tolgod. It gave scientists important clues about how birds and dinosaurs are related.

Perfect Fossils— A Rare Find

Spectacular fossils formed when ancient organisms were buried quickly.

The key to a well-preserved fossil is what happens to an organism right after it dies. When organisms are buried swiftly, their remains are protected from scavengers and natural events. Over time, their bones and teeth form amazingly well-preserved fossils.

As a boy growing up in Los Angeles, Michael Novacek loved visiting the La Brea tar pits. The fossils from these tar pits are remarkably intact because animals became stuck in gooey tar puddles and quickly became submerged. These fossils inspired Novacek to become a paleontologist.

Years later, on an expedition for the American Museum of Natural History, Novacek discovered another extraordinary site. In the Gobi Desert, in Mongolia, Novacek and his team uncovered a rich collection of fossils at a site called Ukhaa Tolgod.

The fossils were astonishingly complete. One fossil was of a 2-inch mammal skeleton, still with its microscopic ear bones. Like the animals at the La Brea tar pits, those that died here were buried quickly. But after examining the evidence, scientists have determined that these animals were likely killed and covered by a devastating avalanche or landslide. Spectacular fossils formed when ancient organisms were buried quickly.

Discovering What Happened at Ukhaa Tolgod

Scientists have looked for clues in the rocks where the fossils were found. Most of the rocks are made of sandstone. One hypothesis was that the animals were buried alive by drifting dunes during a sandstorm. But then scientists noticed the rocks near the fossils held large pebbles that were too big to be carried by wind.

To find an explanation, they turned to Nebraska's Sand Hills—a region with giant, stable dunes similar to those that existed at Ukhaa Tolgod. In the Sand Hills, heavy rains can set off avalanches of wet sand. An enormous slab of heavy, wet sand can bury everything in its path. The current hypothesis about Ukhaa Tolgod is that heavy rains triggered an avalanche of sand that slid down the dunes and buried the animals below.

◄ Novacek and his team have found amazing fossils in Mongolia's Gobi Desert, including this velociraptor.

Today, the dunes at Ukhaa Tolgod are sandy and barren. But long ago, many plants and animals lived in the region. ►

It's Your Turn

DRAW With a partner, draw a comic strip showing how animals in Ukhaa Tolgod might have died and been buried. Use the comic strip to explain how the almost-perfect fossils were preserved.

AMERICAN MUSEUM OF NATURAL HISTORY

Reading Guide

Key Concepts 🔑
ESSENTIAL QUESTIONS

- What does relative age mean?

- How can the positions of rock layers be used to determine the relative ages of rocks?

Vocabulary

relative age p. 337

superposition p. 338

inclusion p. 339

unconformity p. 340

correlation p. 340

index fossil p. 341

g **Multilingual eGlossary**

Relative-Age Dating

Inquiry How did this happen?

Hundreds of millions of years ago, hot magma from deep in Earth was forced into these red, horizontal rock layers in the Grand Canyon. As the magma cooled, it formed this dark gash. How do you think features such as this help scientists determine the relative ages of rock layers?

Which rock layer is oldest?

Scientists study rock layers to learn about the geologic history of an area. How do scientists determine the order in which layers of rock were deposited?

1. Read and complete a lab safety form.

2. Break a **disposable polystyrene meat tray** in half. Place the two pieces on a flat surface so that the broken edges touch one another.

3. Break **another meat tray** in half. Place the two pieces directly on top of the first broken meat tray.

4. Place a **third, unbroken meat tray** on top of the two broken meat trays.

Think About This

1. If you observed rock layers that looked like your model, what would you think might have caused the break only in the two bottom layers?

2. 🔑 **Key Concept** How do you think your model resembles a rock formation? Which layer in your model is youngest? Which is oldest?

Relative Ages of Rocks

You just remembered where you left the money you have been looking for. It is in the pocket of the pants you wore to the movies last Saturday. Look at your pile of dirty clothes. How can you tell where your money is? There really is some order in that pile of dirty clothes. Every time you add clothes to the pile, you place them on top, like the clothes you wore last night. And the clothes from last Saturday are on the bottom. That's where your money is.

Just as there is order in a pile of clothes, there is order in a rock formation. In the rock formation shown in **Figure 9,** the oldest rocks are in the bottom layer and the youngest rocks are in the top layer.

Maybe you have brothers and sisters. If you do, you might describe your age by saying, "I'm older than my sister and younger than my brother." In this way, you compare your age to others in your family. Geologists—the scientists who study Earth and rocks—have developed a set of principles to compare the ages of rock layers. They use these principles to organize the layers according to their relative ages. **Relative age** *is the age of rocks and geologic features compared with other rocks and features nearby.*

🔑 **Key Concept Check** How might you define your relative age?

Figure 9 Just as there is order in a pile of clothes, there is order in this rock formation.

Figure 10 Geologic principles help scientists determine the relative order of rock layers.

✓ **Visual Check** Which rock layer is the oldest?

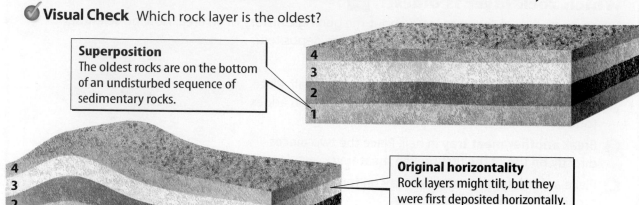

Superposition
The oldest rocks are on the bottom of an undisturbed sequence of sedimentary rocks.

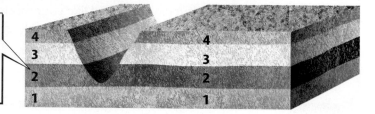

Original horizontality
Rock layers might tilt, but they were first deposited horizontally.

Lateral continuity
Layers are deposited in continuous sheets in all directions until they thin out or hit a barrier. A river might cut through the layers, but the order of layers does not change.

FOLDABLES®

Make a five-tab book and label it as shown. Use it to organize information about the principles of relative-age dating.

Superposition
Original Horizontality
Lateral Continuity
Cross-cutting Relationships
Inclusions

WORD ORIGIN · · · · · · · · · · ·

lateral
from Latin *lateralis*, means "belonging to the side"

Superposition

Your pile of dirty clothes demonstrates the first principle of relative-age dating—superposition. **Superposition** *is the principle that in undisturbed rock layers, the oldest rocks are on the bottom.* Unless some force disturbs the layers after they were deposited, each layer of rocks is younger than the layer below it, as shown in **Figure 10.**

Original Horizontality

An example of the second principle of relative-age dating—original horizontality—is also shown in **Figure 10.** According to the principle of original horizontality, most rock-forming materials are deposited in horizontal layers. Sometimes rock layers are deformed or disturbed after they form. For example, the layers might be tilted or folded. Even though they might be tilted, all the layers were originally deposited horizontally.

✓ **Reading Check** How might rock layers be disturbed?

Lateral Continuity

Another principle of relative-age dating is that sediments are deposited in large, continuous sheets in all **lateral** directions. The sheets, or layers, continue until they thin out or meet a barrier. This principle, called the principle of lateral continuity, is illustrated in the bottom image of **Figure 10.** A river might erode the layers, but their placements do not change.

Sedimentary rock layers

Dike

Inclusions

Fault

1. Sediments are deposited in layers. Eventually, they become layers of rock.

2. Magma intrudes into the rock layers, forming a dike. The dike contains inclusions from the rock layers. The inclusions are older than the dike.

3. Finally, a fault cuts across the rock layers and the dike. The dike is older than the fault, but younger than the rock layers.

Figure 11 Dikes and faults help scientists determine the order in which rock layers were deposited.

Inclusions

Occasionally when rocks form, they contain pieces of other rocks. This can happen when part of an existing rock breaks off and falls into soft sediment or flowing magma. When the sediment or magma becomes rock, the broken piece becomes a part of it. *A piece of an older rock that becomes part of a new rock is called an* **inclusion**. According to the principle of inclusions, if one rock contains pieces of another rock, the rock containing the pieces is younger than the pieces. The vertical intrusion in **Figure 11,** called a dike, is younger than the pieces of rock inside it.

Cross-Cutting Relationships

Sometimes, forces within Earth cause rock formations to break, or fracture. When rocks move along a fracture line, the fracture is called a fault. Faults and dikes cut across existing rock. According to the principle of cross-cutting relationships, if one geologic feature cuts across another feature, the feature that it cuts across is older, as shown in **Figure 11**. This principle is illustrated in the photo at the beginning of this lesson. The black rock layer formed as magma cut across pre-existing red rock layers and crystallized.

 Key Concept Check What geologic principles are used in relative-age dating?

 **MiniLab**   **20 minutes**

Can you model rock layers?

Can a classmate determine the order of your three-dimensional model of rock layers?

1. Read and complete a lab safety form.

2. Cut out a **cube template** as instructed by your teacher.

3. On the sides and top, use **colored pencils** to draw a rock formation that contains 4–5 layers. Include faults, dikes, inclusions, and other disturbances.

4. **Glue** your cube to make a three-dimensional model.

5. Exchange models with another student and determine the order of the layers.

Analyze and Conclude

Key Concept Summarize how positions of rock layers can be used to determine the relative ages of rocks.

Unconformities

After rocks form, they are sometimes uplifted and exposed at Earth's surface. When rocks are exposed, wind and rain start to weather and erode them. These eroded areas represent a gap in the rock record.

Often, new rock layers are deposited on top of old, eroded rock layers. When this happens, an unconformity (un kun FOR muh tee) occurs. *An* **unconformity** *is a surface where rock has eroded away, producing a break, or gap, in the rock record.*

An unconformity is not a hollow gap in the rock. It is a surface on a layer of eroded rocks where younger rocks have been deposited. However, an unconformity does represent a gap in time. It could represent a few hundred years, a million years, or even billions of years. Three major types of unconformities are shown in **Table 1**.

 Key Concept Check How does an unconformity represent a gap in time?

Correlation

You have read that rock layers contain clues about Earth. Geologists use these clues to build a record of Earth's geologic history. Many times the rock record is incomplete, such as happens in an unconformity. Geologists fill in gaps in the rock record by matching rock layers or fossils from separate locations. *Matching rocks and fossils from separate locations is called* **correlation** (kor uh LAY shun).

Matching Rock Layers

Another word for correlation is *connection*. Sometimes it is possible to connect rock layers simply by walking along rock formations and looking for similarities. At other times, soil might cover the rocks, or rocks might be eroded away. In these cases, geologists correlate rocks by matching exposed rock layers in different locations. Through correlation, geologists have established a historical record for part of the southwestern United States, as shown in **Figure 12.**

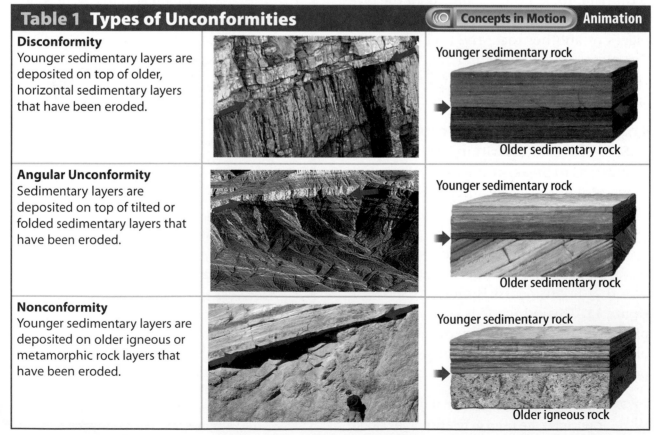

Table 1 Types of Unconformities   Concepts in Motion Animation

Disconformity Younger sedimentary layers are deposited on top of older, horizontal sedimentary layers that have been eroded.		Younger sedimentary rock Older sedimentary rock
Angular Unconformity Sedimentary layers are deposited on top of tilted or folded sedimentary layers that have been eroded.		Younger sedimentary rock Older sedimentary rock
Nonconformity Younger sedimentary layers are deposited on older igneous or metamorphic rock layers that have been eroded.		Younger sedimentary rock Older igneous rock

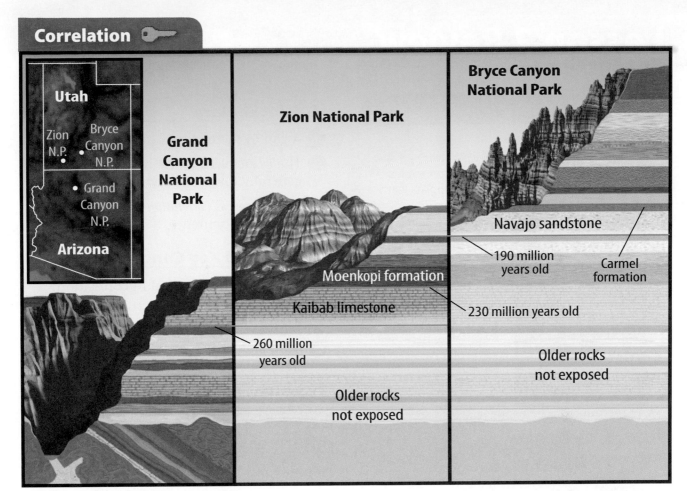

Bryce Canyon National Park

Zion National Park

Grand Canyon National Park

Utah
Zion N.P.
Bryce Canyon N.P.
Grand Canyon N.P.
Arizona

Navajo sandstone

190 million years old

Carmel formation

Moenkopi formation

Kaibab limestone

230 million years old

260 million years old

Older rocks not exposed

Older rocks not exposed

Figure 12 Exposed rock layers from three national parks have been correlated to make a historical record.

 **Visual Check** Which geologic principles must be assumed in order to correlate these layers?

Review
Personal Tutor

Index Fossils

The rock formations in **Figure 12** are correlated based on similarities in rock type, structure, and fossil evidence. They exist within a few hundred kilometers of one another. If scientists want to learn the relative ages of rock formations that are very far apart or on different continents, they often use fossils. If two or more rock formations contain fossils of about the same age, scientists can infer that the formations are also about the same age.

Not all fossils are useful in determining the relative ages of rock layers. Fossils of species that lived on Earth for hundreds of millions of years are not helpful. They represent time spans that are too long. The most useful fossils represent species, like certain trilobites, that existed for only a short time in many different areas on Earth. These fossils are called index fossils. **Index fossils** *represent species that existed on Earth for a short length of time, were abundant, and inhabited many locations.* When an index fossil is found in rock layers at different locations, geologists can infer that the layers are of similar age.

Key Concept Check How are index fossils useful in relative-age dating?

Lesson 2 Review

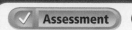

Visual Summary

Geologic principles help geologists learn the relative ages of rock layers.

The rock record is incomplete because some of it has eroded away.

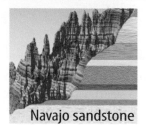

Navajo sandstone

Geologists fill in gaps in the rock record by correlating rock layers.

FOLDABLES®

Use your lesson Foldable to review the lesson. Save your Foldable for the project at the end of the chapter.

What do you think NOW?

You first read the statements below at the beginning of the chapter.

3. Older rocks are always located below younger rocks.

4. Relative age means that scientists are relatively sure of the age.

Did you change your mind about whether you agree or disagree with the statements? Rewrite any false statements to make them true.

Use Vocabulary

1 A gap in the rock record is a(n) _____.

2 The principle that the oldest rocks are generally on the bottom is _____.

3 **Use the terms** *correlation* and *index fossil* in a complete sentence.

Understand Key Concepts

4 Which might be useful in correlation?
   A. amber   C. trilobite
   B. inclusion   D. unconformity

5 **Draw** and label a sequence of rock layers showing how an unconformity might form.

6 **Relate** uniformitarianism to principles of relative-age dating.

Interpret Graphics

Use the diagram below to answer question 7.

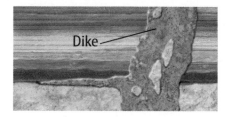

Dike

7 **Decide** Which is older—the rock layers or the dike? Explain which geologic principle you used to arrive at your answer.

8 **Summarize** Copy and fill in the graphic organizer below to identify five geologic principles useful in relative-age dating.

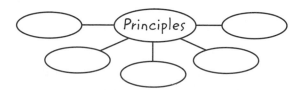

Principles

Critical Thinking

9 **Evaluate** why fossils might be more useful than rock types in correlating rock layers on two different continents.

10 **Debate** whether you think humans might be useful as index fossils in the future.

Can you correlate rock formations?

Most rocks have been buried in Earth for thousands, millions, or even billions of years. Occasionally, rock layers become exposed on Earth's surface. To correlate rock layers exposed at different locations, it is sometimes necessary to **interpret scientific illustrations** of the layers.

Materials

pencil

colored pencils

large, soft eraser

ruler

Learn It

Drawings and photos can make complex scientific data easier to understand. Use the drawings below to represent rock formations. As you correlate the layers, use the key to **interpret each illustration.**

Try It

1 As well as you can, copy the drawings of the four rock columns shown below into your Science Journal. *Do not write in this book.*

2 Color your drawings so that each rock layer is one color in each of the four rock columns. Use the key to determine what type of rocks each layer contains.

3 Carefully study your drawings. Try to determine which rock columns correlate the best.

Apply It

4 Which rock columns correlated the best? Which principle of relative-age dating did you use when correlating the rock layers?

5 Which rock layer is the oldest in rock column C? The youngest? Which geologic principle did you use to determine this?

6 Identify the type of unconformity that exists in rock column B.

7 🗝 **Key Concept** How can you use types of rocks to correlate rock layers? What other type of evidence could you use to determine the relative ages of rock layers?

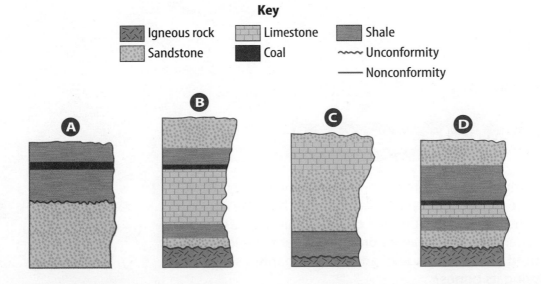

Key

⬚ Igneous rock   ⬚ Limestone   ⬚ Shale
⬚ Sandstone   ⬚ Coal   ∿ Unconformity
— Nonconformity

Key Concepts 🔑
ESSENTIAL QUESTIONS

- What does absolute age mean?
- How can radioactive decay be used to date rocks?

Vocabulary

absolute age p. 345

isotope p. 346

radioactive decay p. 346

half-life p. 347

g **Multilingual eGlossary**

Absolute-Age Dating

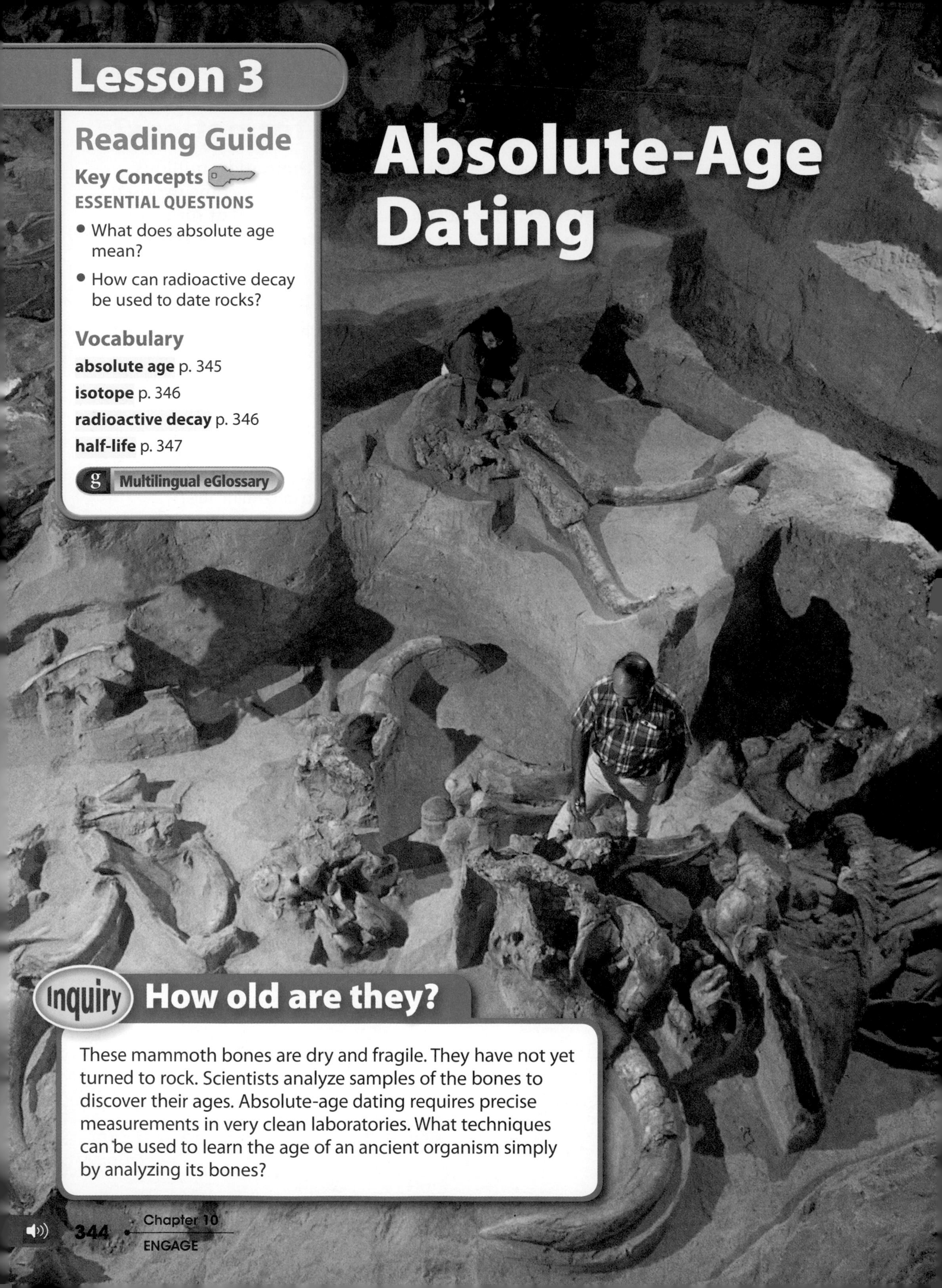

Inquiry How old are they?

These mammoth bones are dry and fragile. They have not yet turned to rock. Scientists analyze samples of the bones to discover their ages. Absolute-age dating requires precise measurements in very clean laboratories. What techniques can be used to learn the age of an ancient organism simply by analyzing its bones?

How can you describe your age?

If you described your relative age compared to your classmates', how would you do it? How do you think your actual, or absolute, age differs from your relative age?

1. One student will write down his or her birth date on an **index card.** The student will hold the card while everyone else files by and looks at it.

2. Form two groups depending on whether your birth date falls before or after the date on the card.

3. Remaining in your group, write down your own birth date on an index card. Quietly form a line in order of your birth dates.

Think About This

1. When you were in two groups, what did you know about everyone's age? When you lined up, what did you know about everyone's age? Which is your relative age? Your absolute age?

2. Can you think of a situation where it would be important to know your absolute age?

3. 🔑 **Key Concept** Why do you think scientists would want to know the absolute age of a rock?

Absolute Ages of Rocks

Recall from Lesson 2 that you have a relative age. You might be older than your sister and younger than your brother—or you might be the youngest in your family. You also can describe your age by saying your age in years, such as "I am 13 years old." This is not a relative age. It is your age in numbers—your numerical age.

Similarly, scientists can describe the ages of some kinds of rocks numerically. Scientists use the term **absolute age** *to mean the numerical age, in years, of a rock or object.* By measuring the absolute ages of rocks, geologists have developed accurate historical records for many geologic formations.

🔑 **Key Concept Check** How is absolute age different from relative age?

Scientists have been able to determine the absolute ages of rocks and other objects only since the beginning of the twentieth century. That is when radioactivity was discovered. Radioactivity is the release of energy from unstable atoms. The image in **Figure 13** was made using X-rays. How can radioactivity be used to date rocks? In order to answer this question, you need to know about the internal structure of the atoms that make up elements.

Figure 13 The release of radioactive energy can be used to make an X-ray.

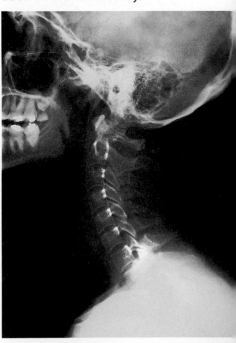

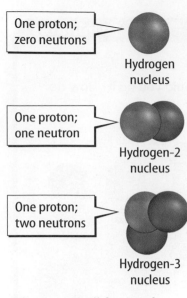

One proton;
zero neutrons

Hydrogen
nucleus

One proton;
one neutron

Hydrogen-2
nucleus

One proton;
two neutrons

Hydrogen-3
nucleus

Figure 14 All forms of hydrogen contain only one proton regardless of the number of neutrons.

WORD ORIGIN ············

isotope
from Greek *isos*, means "equal"; and *topos*, means "place"

Atoms

You are probably familiar with the periodic table of the elements, which is shown inside the back cover of this book. Each element is made up of atoms. An atom is the smallest part of an element that has all the properties of the element. Each atom contains smaller particles called protons, neutrons, and electrons. Protons and neutrons are in an atom's nucleus. Electrons surround the nucleus.

Isotopes

All atoms of a given element have the same number of protons. For example, all hydrogen atoms have one proton. But an element's atoms can have different numbers of neutrons. The three atoms shown in **Figure 14** are all hydrogen atoms. Each has the same number of protons—one. However, one of the hydrogen atoms has no neutrons, one has one neutron, and the other has two neutrons. The three different forms of hydrogen atoms are called hydrogen **isotopes** (I suh tohps). **Isotopes** *are atoms of the same element that have different numbers of neutrons.*

 Reading Check How do an element's isotopes differ?

Radioactive Decay

Most isotopes are stable. Stable isotopes do not change under normal conditions. But some isotopes are unstable. These isotopes are known as radioactive isotopes. Radioactive isotopes decay, or change, over time. As they decay, they release energy and form new, stable atoms. **Radioactive decay** *is the process by which an unstable element naturally changes into another element that is stable.* The unstable isotope that decays is called the parent isotope. The new element that forms is called the daughter isotope. **Figure 15** illustrates an example of radioactive decay. The atoms of an unstable isotope of hydrogen (parent) decay into atoms of a stable isotope of helium (daughter).

Radioactive Decay

Figure 15 An unstable parent hydrogen isotope produces the stable daughter helium isotope.

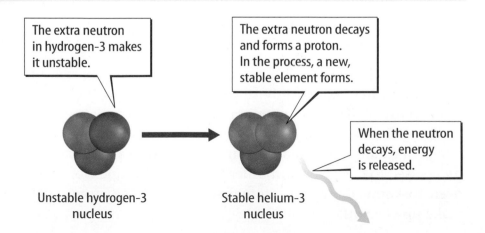

The extra neutron in hydrogen-3 makes it unstable.

The extra neutron decays and forms a proton. In the process, a new, stable element forms.

When the neutron decays, energy is released.

Unstable hydrogen-3 nucleus

Stable helium-3 nucleus

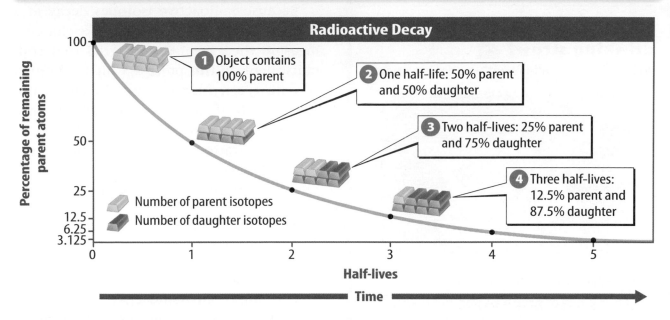

Figure 16 The half-life is the time it takes for one-half of the parent isotopes to change into daughter isotopes.

Visual Check What percentages of parent isotopes and daughter isotopes will there be after four half-lives?

Half-Life

The rate of decay from parent isotopes into daughter isotopes is different for different radioactive elements. But the rate of decay is constant for a given isotope. This rate is measured in time units called half-lives. *An isotope's* **half-life** *is the time required for half of the parent isotopes to decay into daughter isotopes.* Half-lives of radioactive isotopes range from a few microseconds to billions of years.

Reading Check What is half-life?

The graph in **Figure 16** shows how half-life is measured. As time passes, more and more unstable parent isotopes decay and form stable daughter isotopes. That means the ratio between the numbers of parent and daughter isotopes is always changing. When half the parent isotopes have decayed into daughter isotopes, the isotope has reached one half-life. At this point, 50 percent of the isotopes are parents and 50 percent of the isotopes are daughters. After two half-lives, one-half of the remaining parent isotopes have decayed so that only one-quarter as much parent remains as at the start. At this point, 25 percent of the isotopes are parent and 75 percent of the isotopes are daughter. After three half-lives, half again of the remaining parent isotopes have decayed into daughter isotopes. This process continues until nearly all parent isotopes have decayed into daughter isotopes.

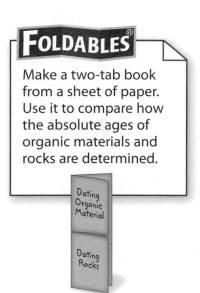

FOLDABLES

Make a two-tab book from a sheet of paper. Use it to compare how the absolute ages of organic materials and rocks are determined.

Inquiry MiniLab
10 minutes

What is the half-life of a drinking straw?

You can model half-life with a drinking straw.

1 Read and complete a lab safety form.

2 On a piece of **graph paper,** draw an *x*-axis and a *y*-axis. Label the *x*-axis *Number of half-lives,* from 0 to 4 in equal intervals. Leave the *y*-axis blank.

3 Use a **metric ruler** to measure a **drinking straw.** Mark its height on the *y*-axis, as shown in the photo. Use **scissors** to cut the straw in half and discard half of it. Mark the height of the remaining half as the first half-life.

4 Repeat four times, each time cutting the straw in half and each time adding a measurement to your graph's *y*-axis.

Analyze and Conclude

1. Compare your graph to the graph in **Figure 16.** How is it similar? How is it different?

2. 🔑 **Key Concept** Explain how your disappearing straw represents the decay of a radioactive element.

Radiometric Ages

Because radioactive isotopes decay at a constant rate, they can be used like clocks to measure the age of the material that contains them. In this process, called radiometric dating, scientists measure the amount of parent isotope and daughter isotope in a sample of the material they want to date. From this ratio, they can determine the material's age. Scientists make these very precise measurements in laboratories.

✓ **Reading Check** What is measured in radiometric dating?

Radiocarbon Dating

One important radioactive isotope used for dating is an isotope of carbon called radiocarbon. Radiocarbon is also known as carbon-14, or C-14, because there are 14 particles in its nucleus—six protons and eight neutrons. Radiocarbon forms in Earth's upper atmosphere. There, it mixes in with a stable isotope of carbon called carbon-12, or C-12. The ratio of C-14 to C-12 in the atmosphere is constant.

All living things use carbon as they build and repair tissues. As long as an organism is alive, the ratio of C-14 to C-12 in its tissues is identical to the ratio in the atmosphere. However, when an organism dies, it stops taking in C-14. The C-14 already present in the organism starts to decay to nitrogen-14 (N-14). As the dead organism's C-14 decays, the ratio of C-14 to C-12 changes. Scientists measure the ratio of C-14 to C-12 in the remains of the dead organism to determine how much time has passed since the organism died.

The half-life of carbon-14 is 5,730 years. That means radiocarbon dating is useful for measuring the age of the remains of organisms that died up to about 60,000 years ago. In older remains, there is not enough C-14 left to measure accurately. Too much of it has decayed to N-14.

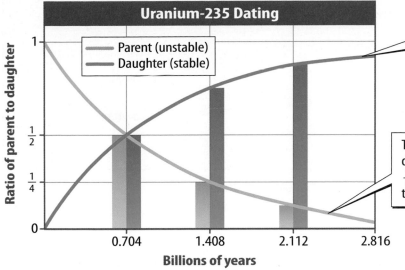

Uranium-235 Dating

Ratio of parent to daughter

- Parent (unstable)
- Daughter (stable)

0.704   1.408   2.112   2.816

Billions of years

An unstable parent isotope (U-235) will decay at a constant rate and form a daughter product (Pb-207). After one half-life, the concentrations of parent and daughter isotopes are equal.

The parent isotope will continue to decay over time. After two half-lives, $\frac{1}{4}$ of the original parent remains. After three half-lives, $\frac{1}{8}$ remains, and so on.

Figure 17 Scientists determine the absolute age of an igneous rock by measuring the ratio of uranium-235 isotopes (parent) to lead-207 isotopes (daughter) in the rock's minerals.

Visual Check How old is a mineral that contains 25 percent U-235?

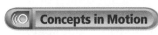

Animation

REVIEW VOCABULARY

mineral
a naturally occurring, inorganic solid with a definite chemical composition and an orderly arrangement of atoms

Dating Rocks

Radiocarbon dating is useful only for dating organic material—material from once-living organisms. This material includes bones, wood, parchment, and charcoal. Most rocks do not contain organic material. Even most fossils are no longer organic. In most fossils, living tissue has been replaced by rock-forming **minerals.** For dating rocks, geologists use different kinds of radioactive isotopes.

Dating Igneous Rock One of the most common isotopes used in radiometric dating is uranium-235, or U-235. U-235 is often trapped in the minerals of igneous rocks that crystallize from hot, molten magma. As soon as U-235 is trapped in a mineral, it begins to decay to lead-207, or Pb-207, as shown in **Figure 17.** Scientists measure the ratio of U-235 to Pb-207 in a mineral to determine how much time has passed since the mineral formed. This provides the age of the rock that contains the mineral.

Dating Sedimentary Rock In order to be dated by radiometric means, a rock must have U-235 or other radioactive isotopes trapped inside it. The grains in many sedimentary rocks come from a variety of weathered rocks from different locations. The radioactive isotopes within these grains generally record the ages of the grains—not the time when the sediment was deposited. For this reason, sedimentary rock is not as easily dated as igneous rock in radiometric dating.

Key Concept Check Why are radioactive isotopes not useful for dating sedimentary rocks?

Table 2 Radioactive Isotopes Used for Dating Rocks

Parent Isotope	Half-Life	Daughter Product
Uranium-235	704 million years	lead-207
Potassium-40	1.25 billion years	argon-40
Uranium-238	4.5 billion years	lead-206
Thorium-232	14.0 billion years	lead-208
Rubidium-87	48.8 billion years	strontium-87

Table 2 Radioactive isotopes useful for dating rocks have long half-lives.

Math Skills

Use Significant Digits

The answer to a problem involving measurement cannot be more precise than the measurement with the fewest number of significant digits. For example, if you begin with 36 grams (2 significant digits) of U-235, how much U-235 will remain after 2 half-lives?

1. After the first half-life, $\frac{36\ g}{2} = 18$ g of U-235 remain.

2. After the second half-life, $\frac{18\ g}{2} = 9.0$ g of U-235 remain. Add the zero to retain two significant digits.

Practice

The half-life of rubidium-87 (Rb-87) is 48.8 billion years. What is the length of three half-lives of Rb-87?

Review

- **Math Practice**
- **Personal Tutor**

Different Types of Isotopes The half-life of uranium-235 is 704 million years. This makes it useful for dating rocks that are very old. **Table 2** lists five of the most useful radioactive isotopes for dating old rocks. All of them have long half-lives. Radioactive isotopes with short half-lives cannot be used for dating old rocks. They do not contain enough parent isotope to measure. Geologists often use a combination of radioactive isotopes to measure the age of a rock. This helps make the measurements more accurate.

Key Concept Check Why is a radioactive isotope with a long half-life useful in dating very old rocks?

The Age of Earth

The oldest known rock formation dated by geologists using radiometric means is in Canada. It is estimated to be between 4.03 billion and 4.28 billion years old. However, individual crystals of the mineral zircon in igneous rocks in Australia have been dated at 4.4 billion years.

With rocks and minerals more than 4 billion years old, scientists know that Earth must be at least that old. Radiometric dating of rocks from the Moon and meteorites indicate that Earth is 4.54 billion years old. Scientists accept this age because evidence suggests that Earth, the Moon, and meteorites all formed at about the same time.

Radiometric dating, the relative order of rock layers, and fossils all help scientists understand Earth's long history. Understanding Earth's history can help scientists understand changes occurring on Earth today—as well as changes that are likely to occur in the future.

Visual Summary

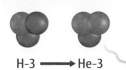

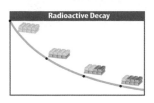

When the unstable atoms of radioactive isotopes decay, they form new, stable isotopes.

H-3 ⟶ He-3

Because radioactive isotopes decay at constant rates, they can be used to determine absolute ages.

Radioactive Decay

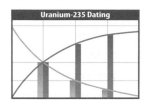

Uranium-235 Dating

Isotopes with long half-lives are the most useful for dating old rocks.

FOLDABLES

Use your lesson Foldable to review the lesson. Save your Foldable for the project at the end of the chapter.

What do you think NOW?

You first read the statements below at the beginning of the chapter.

5. Absolute age means that scientists are sure of the age.

6. Scientists use radioactive decay to determine the ages of some rocks.

Did you change your mind about whether you agree or disagree with the statements? Rewrite any false statements to make them true.

Use Vocabulary

1 **Compare** absolute age and relative age.

2 The rate of radioactive decay is expressed as an isotope's _____.

3 **Use the terms** *atom* and *isotope* in a complete sentence.

Understand Key Concepts

4 Which could you date with carbon-14?
  A. a fossilized shark's tooth
  B. an arrowhead carved out of rock
  C. a petrified tree
  D. charcoal from an ancient campfire

5 **Explain** why radioactive isotopes are more useful for dating igneous rocks than they are for dating sedimentary rocks.

6 **Differentiate** between parent isotopes and daughter isotopes.

Interpret Graphics

7 **Identify** Copy and fill in the graphic organizer below to identify the three parts of an atom.

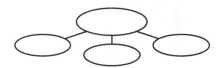

Critical Thinking

8 **Evaluate** the importance of radioactive isotopes in determining the age of Earth.

Math Skills
─── Math Practice ───

9 The half life of potassium-40 (K-40) is 1.25 billion years. If you begin with 130 g of K-40, how much remains after 2.5 billion years? Use the correct number of significant digits in your answer.

Correlate Rocks Using Index Fossils

Imagine you are a geologist and you have been asked to correlate the rock columns below in order to determine the relative ages of the layers. Recall that geologists can correlate rock layers in different ways. In this lab, use index fossils to correlate and date the layers.

Question

How can index fossils be used to determine the relative ages of Earth's rocks?

Procedure

1. Carefully examine the three rock columns on this page. Each rock layer can be identified with a letter and a number. For example, the second layer down in column A is layer A-2.

2. In your Science Journal, correlate the layers using only the fossils—not the types of rock. Before you begin, look at the fossil key on the next page. It shows the time intervals during which each organism or group of organisms lived on Earth. Refer to the key as you correlate.

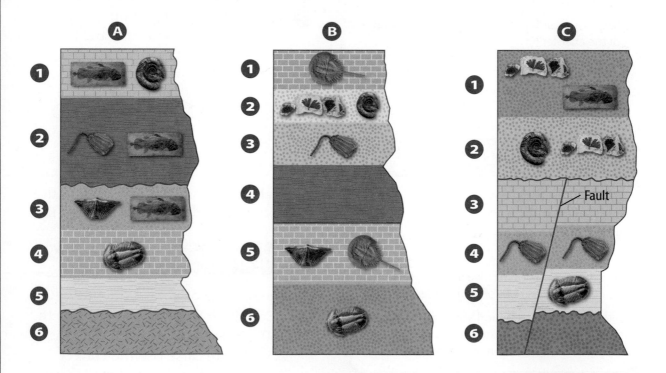

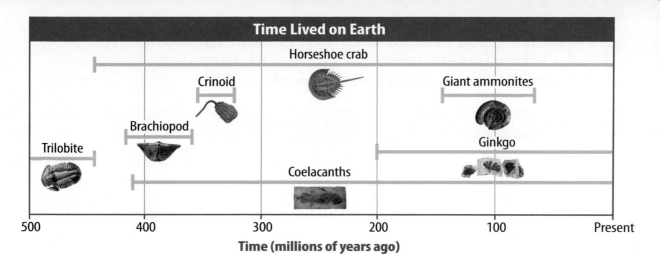

Time Lived on Earth

Horseshoe crab

Crinoid

Giant ammonites

Brachiopod

Ginkgo

Trilobite

Coelacanths

| 500 | 400 | 300 | 200 | 100 | Present |

Time (millions of years ago)

Analyze and Conclude

3 **Differentiate** Which fossils in the key appear to be index fossils? Explain your choices.

4 **Match** Correlate layer A-2 to one layer in each of the other two columns. Approximately how old are these layers? How do you know?

5 **Infer** What is the approximate age of layer B-4? *Hint: It lies between two index fossils.*

6 **Infer** How old is the fault in column C?

7 **Compare and Contrast** How does correlating rocks using fossils differ from correlating rocks using types of rock?

8 **BIG** **The Big Idea** How can fossils be used to determine the relative ages of rocks?

Communicate Your Results

Choose a partner. One of you is a reporter and one is a geologist. Conduct an interview about what kinds of fossils are best used to date rocks.

 **Extension**

Choose one of the three rock formations you correlated. Based on your results, provide a range of dates for each of the layers within it.

Lab Tips

☑ You might want to copy the rock layers in your Science Journal and correlate them by drawing lines connecting the layers.

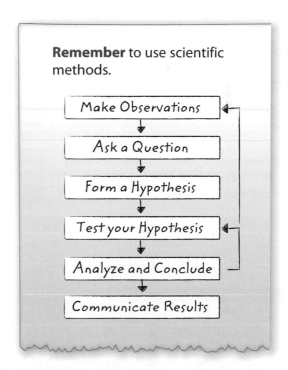

Remember to use scientific methods.

Make Observations

↓

Ask a Question

↓

Form a Hypothesis

↓

Test your Hypothesis

↓

Analyze and Conclude

↓

Communicate Results

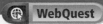
THE BIG IDEA
Evidence from fossils, rock layers, and radioactivity help scientists understand Earth's history and determine the ages of Earth's rocks.

Key Concepts Summary 🔑	Vocabulary
Lesson 1: Fossils • A **fossil** is the preserved remains or evidence of ancient organisms. Organisms are more likely to become fossils if they have hard parts and are buried quickly after they die. Fossils include **carbon films**, **molds**, **casts**, and **trace fossils**. • **Paleontologists** use clues from fossils to learn about ancient life and the environments ancient organisms lived in. 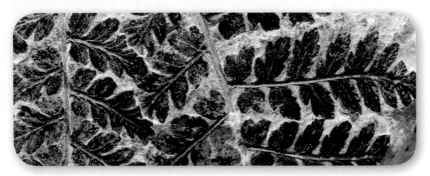	**fossil** p. 327 **catastrophism** p. 327 **uniformitarianism** p. 328 **carbon film** p. 330 **mold** p. 331 **cast** p. 331 **trace fossil** p. 331 **paleontologist** p. 332
Lesson 2: Relative-Age Dating • **Relative age** is the age of rocks and geologic features compared with rocks and features nearby. • The relative age of rock layers can be determined using geologic principles, such as the principle of **superposition** and the principle of **inclusion**. **Unconformities** represent time gaps in the rock record. 	**relative age** p. 337 **superposition** p. 338 **inclusion** p. 339 **unconformity** p. 340 **correlation** p. 340 **index fossil** p. 341
Lesson 3: Absolute-Age Dating 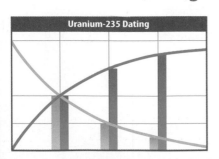 • **Absolute age** is the age in years of a rock or object. • The **radioactive decay** of unstable **isotopes** occurs at a constant rate, measured as **half-life.** To date a rock or object, scientists measure the ratios of its parent and daughter isotopes.	**absolute age** p. 345 **isotope** p. 346 **radioactive decay** p. 346 **half-life** p. 347

FOLDABLES® **Chapter Project**

Assemble your lesson Foldables® as shown to make a Chapter Project. Use the project to review what you have learned in this chapter.

Fossils

Superposition
Original Horizontality
Lateral Continuity
Cross-cutting Relationships
Inclusions

Dating Organic Material

Dating Rocks

Use Vocabulary

1. An ancient dinosaur track is a(n) _____ .

2. _____ use the principle of _____ to reconstruct ancient environments.

3. The principle of _____ states that the oldest layers are generally at the bottom.

4. In _____ , geologists use _____ to match rock layers on separate continents.

5. A(n) _____ is an eroded surface.

6. The process of _____ can be used like a clock to determine a rock's _____ .

7. A Uranium-235 _____ decays with a constant _____ of 704 million years.

Link Vocabulary and Key Concepts

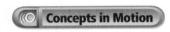 **Concepts in Motion** **Interactive Concept Map**

Copy this concept map, and then use vocabulary terms from the previous page to complete the concept map.

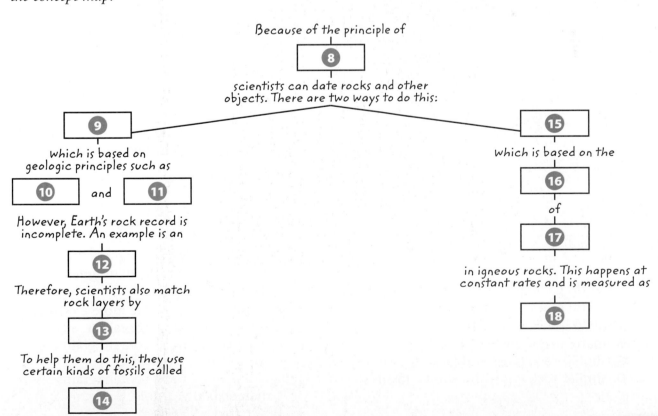

Because of the principle of

8

scientists can date rocks and other objects. There are two ways to do this:

9

which is based on geologic principles such as

10 and **11**

However, Earth's rock record is incomplete. An example is an

12

Therefore, scientists also match rock layers by

13

To help them do this, they use certain kinds of fossils called

14

15

which is based on the

16

of

17

in igneous rocks. This happens at constant rates and is measured as

18

Understand Key Concepts 🗝

1 Which idea explains Earth's history by examining present conditions on Earth?

A. absolute-age dating
B. catastrophism
C. relative-age dating
D. uniformitarianism

2 Which part of a dinosaur is least likely to be fossilized?

A. bone
B. brain
C. horn
D. tooth

3 Which makes a species a good index fossil?

A. lived a long time and was abundant
B. lived a long time and was scarce
C. lived a short time and was scarce
D. lived a short time and was abundant

4 In the drawing below, what is the order of rock layers from oldest to youngest?

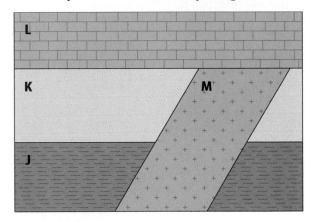

A. J, K, L, M
B. J, K, M, L
C. L, K, J, M
D. M, J, K, L

5 What do geologists look for in order to correlate rocks in different locations?

A. different rock types and similar fossils
B. many rock types and many fossils
C. similar rock types and lack of fossils
D. similar rock types and similar fossils

6 What is the half-life on the graph below?

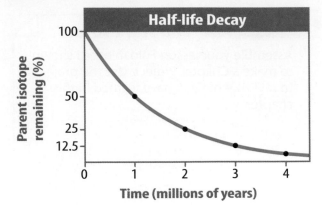

A. 1 million years
B. 2 million years
C. 3 million years
D. 4 million years

7 What are isotopes?

A. atoms of the same element with different numbers of electrons but the same number of protons
B. atoms of the same element with different numbers of electrons but the same number of neutrons
C. atoms of the same element with different numbers of neutrons but the same number of protons
D. atoms of the same element with equal numbers of neutrons and protons.

8 What do scientists measure when determining the absolute age of a rock?

A. amount of radioactivity
B. number of uranium atoms
C. ratio of neutrons and electrons
D. ratio of parent and daughter isotopes

9 Why is radiometric dating less useful to date sedimentary rocks than igneous rocks?

A. Sedimentary rocks are more eroded.
B. Sedimentary rocks contain fossils.
C. Sedimentary rocks contain grains formed from other rocks.
D. Sedimentary rocks contain grains less than 60,000 years old.

Critical Thinking

10 **Give** an example of superposition from your own life.

11 **Suggest** a way that an ancient human might have been preserved as a fossil.

12 **Explain** why scientists use a combination of uniformitarianism and catastrophism ideas to understand Earth.

13 **Reason** You are studying a rock formation that includes layers of folded sedimentary rocks cut by faults and dikes. Describe the geologic principles you would use to determine the relative order of the layers.

14 **Construct** a graph showing the radioactive decay of an unstable isotope with a half-life of 250 years. Label three half-lives.

15 **Assess** The ash layers in the drawing below have been dated as shown. What conclusions can you draw about the ages of each of the layers A, B, and C?

C
Ash deposited 540 mya
B
Ash deposited 730 mya
A

Writing in Science

16 **Write** a paragraph of at least five sentences explaining why absolute-age dating has been more useful than relative-age dating in determining the age of Earth. Include a main idea, supporting details, and concluding sentence.

REVIEW THE B|G IDEA

17 What evidence do scientists use to determine the ages of rocks?

18 The photo below shows many rock layers of the Grand Canyon. Explain how the development of the principle of uniformitarianism might have changed earlier ideas about the age of the Grand Canyon and how it formed.

Math Skills ×÷

Review

Math Practice

Use Significant Figures

19 If you begin with 68 g of an isotope, how many grams of the original isotope will remain after four half-lives?

20 The half-life of radon-222 (Rn-222) is 3.823 days.
a. How long would it take for three half-lives?
b. What percentage of the original sample would remain after three half-lives?

21 The half-life of Rn-222 is 3.823 days. What was the original mass of a sample of this isotope if 0.0500 g remains after 7.646 days?

Record your answers on the answer sheet provided by your teacher or on a sheet of paper.

Multiple Choice

1 Which is a copy of a dead organism formed when its impression fills with mineral deposits or sediments?

   A carbon film

   B cast

   C mold

   D trace fossil

Use the diagram below to answer question 2.

2 In the diagram above, which rock layer typically is youngest?

   A 1

   B 2

   C 3

   D 4

3 Which characteristic of rocks does radioactive decay measure?

   A absolute age

   B lateral continuity

   C relative age

   D unconformity

4 Which increases the likelihood that a dead organism will be fossilized?

   A fast decay of bones

   B presence of few hard body parts

   C quick burial after death

   D vast amounts of skin

Use the diagram below to answer question 5.

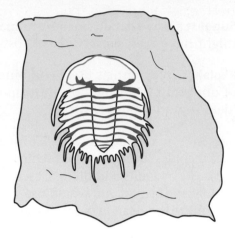

5 Which fossilized ancient organism is pictured in the diagram above?

   A clam

   B mammoth

   C mastodon

   D trilobite

6 Which explains most of Earth's geological features as a result of short periods of earthquakes, volcanoes, and meteorite impacts?

   A catastrophism

   B evolution

   C supernaturalism

   D uniformitarianism

7 Which fossil type helps geologists infer that rock layers in different geographic locations are similar in age?

   A carbon film

   B index fossil

   C preserved remains

   D trace fossil

8 Which pie chart shows the ratio of parent to daughter atoms after four half-lives?

A

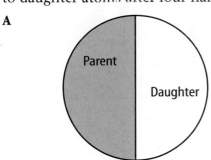

B

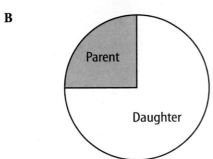

C

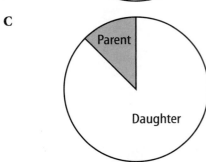

D

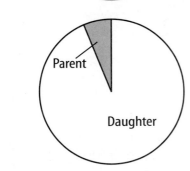

Constructed Response

Use the diagram below to answer questions 9 and 10.

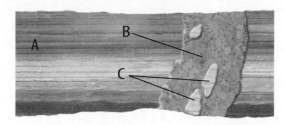

9 Are the sedimentary rock layers (A) older or younger than the dike (B)? How do you know?

10 Is the dike (B) older or younger than the inclusions (C)? How do you know?

Use the diagram below to answer question 11.

Younger sedimentary rock

Older sedimentary rock

11 Identify the type of unconformity that exists in the diagram above. Hypothesize how this could have happened.

12 What is C-14? What role does it play in radiocarbon dating? Why does time limit the effectiveness of radiocarbon dating as a tool for measuring age?

NEED EXTRA HELP?												
If You Missed Question...	1	2	3	4	5	6	7	8	9	10	11	12
Go to Lesson...	1	2	3	1	1	1	2	3	2	2	2	3

Geologic Time

THE BIG IDEA

What have scientists learned about Earth's past by studying rocks and fossils?

Inquiry | What happened to the dinosaurs?

This Triceratops lived millions of years ago. Hundreds of other kinds of dinosaurs lived at the same time. Some were as big as houses; others were as small as chickens. Scientists learn about dinosaurs by studying their fossils. Like many organisms that have lived on Earth, dinosaurs disappeared suddenly. Why did the dinosaurs disappear?

- How has Earth changed over geologic time?
- How do geologic events affect life on Earth?
- What have scientists learned about Earth's past by studying rocks and fossils?

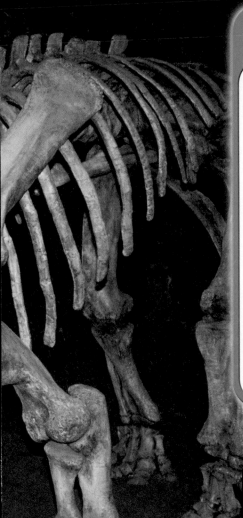

Get Ready to Read

What do you think?

Before you read, decide if you agree or disagree with each of these statements. As you read this chapter, see if you change your mind about any of the statements.

1. All geologic eras are the same length of time.

2. Meteorite impacts cause all extinction events.

3. North America was once on the equator.

4. All of Earth's continents were part of a huge supercontinent 250 million years ago.

5. All large Mesozoic vertebrates were dinosaurs.

6. Dinosaurs disappeared in a large mass extinction event.

7. Mammals evolved after dinosaurs became extinct.

8. Ice covered nearly one-third of Earth's land surface 10,000 years ago.

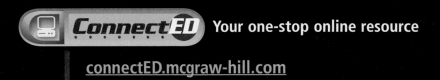

ConnectED Your one-stop online resource

<u>connectED.mcgraw-hill.com</u>

Video

Audio

Review

Inquiry

WebQuest

Assessment

Concepts in Motion

Multilingual eGlossary

Lesson 1

Geologic History and the Evolution of Life

Reading Guide

Key Concepts 🔑
ESSENTIAL QUESTIONS

- How was the geologic time scale developed?
- What are some causes of mass extinctions?
- How is evolution affected by environmental change?

Vocabulary

eon p. 363

era p. 363

period p. 363

epoch p. 363

mass extinction p. 365

land bridge p. 366

geographic isolation p. 366

g Multilingual eGlossary

🎞 Video

- BrainPOP®
- Science Video

Inquiry What happened here?

A meteorite 50 m in diameter crashed into Earth 50,000 years ago. The force of the impact created this crater in Arizona and threw massive amounts of dust and debris into the atmosphere. Scientists hypothesize that a meteorite 200 times this size—the size of a small city—struck Earth 65 million years ago. How might it have affected life on Earth?

Can you make a time line of your life?

How would you organize a time line of your life? You might include regular events, such as birthdays. But you might also include special events, such as a weekend camping trip or a summer vacation.

① Read and complete a lab safety form.

② Use **scissors** to cut two pieces of **graph paper** in half. **Tape** them together to make one long piece of paper. Write down the years of your life in horizontal sequence, marked off at regular intervals.

③ Choose up to 12 important events or periods of time in your life. Mark those events on your time line.

Think About This

1. Do the events on your time line appear at regular intervals?

2. 🔑 **Key Concept** How do you think the geologic time scale is like a time line of your life?

Developing a Geologic Time Line

Think about what you did over the last year. Maybe you went on vacation during the summer or visited relatives in the fall. To organize events in your life, you use different units of time, such as weeks, months, and years. Geologists organize Earth's past in a similar way. They developed a time line of Earth's past called the geologic time scale. As shown in **Figure 1,** time units on the geologic time scale are thousands and millions of years long—much longer than the units you use to organize events in your life.

Units in the Geologic Time Scale

Eons *are the longest units of geologic time.* Earth's current eon, the Phanerozoic (fan er oh ZOH ihk) eon, began 542 million years ago (mya). *Eons are subdivided into smaller units of time called* **eras.** *Eras are subdivided into* **periods.** *Periods are subdivided into* **epochs** (EH pocks). Epochs are not shown on the time line in **Figure 1.** Notice that the time units are not equal. For example, the Paleozoic era is longer than the Mesozoic and Cenozoic eras combined.

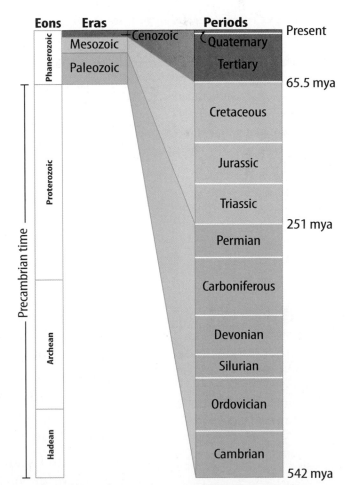

Figure 1 In the geologic time scale, the 4.6 billion years of Earth's history are divided into time units of unequal length.

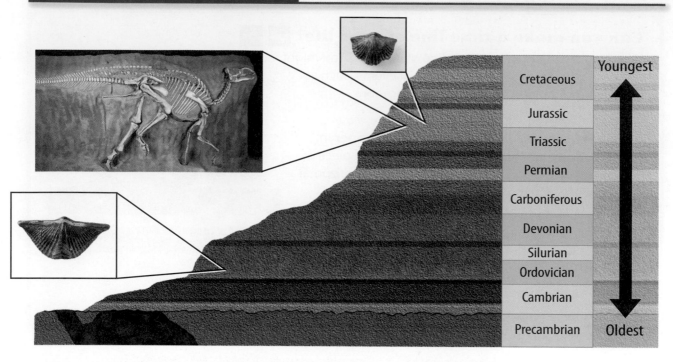

Figure 2 Both older and younger rocks contain fossils of small, relatively simple life-forms. Only younger rocks contain larger, more complex fossils.

SCIENCE USE V. COMMON USE

scale

Science Use a series of marks or points at known intervals

Common Use an instrument used for measuring the weight of an object

The Time Scale and Fossils

Hundreds of years ago, when geologists began developing the geologic time scale, they chose the time boundaries based on what they observed in Earth's rock layers. Different layers contained different fossils. For example, older rocks contained only fossils of small, relatively simple life-forms. Younger rocks contained these fossils as well as fossils of other, more complex organisms, such as dinosaurs, as illustrated in **Figure 2.**

Major Divisions in the Geologic Time Scale

While studying the fossils in rock layers, geologists often saw abrupt changes in the types of fossils within the layers. Sometimes, fossils in one rock layer did not appear in the rock layers right above it. It seemed as though the organisms that lived during that period of time had disappeared suddenly. Geologists used these sudden changes in the fossil record to mark divisions in geologic time. Because the changes did not occur at regular intervals, the boundaries between the units of time in the geologic time scale are irregular. This means the time units are of unequal length.

The time scale is a work in progress. Scientists debate the placement of the boundaries as they make new discoveries.

🔑 **Key Concept Check** Why are fossils important in the development of the geologic time scale?

FOLDABLES

Make a four-door book from a vertical sheet of paper. Use it to organize information about the units of geologic time.

Eon   Era

Period   Epoch

Responses to Change

Sudden changes in the fossil record represent times when large populations of organisms died or became extinct. *A **mass extinction** is the extinction of many species on Earth within a short period of time.* As shown in **Figure 3,** there have been several mass extinction events in Earth's history.

Changes in Climate

What could cause a mass extinction? All species of organisms depend on the environment for their survival. If the environment changes quickly and species do not adapt to the change, they die.

Many things can cause a climate change. For example, gas and dust from volcanoes can block sunlight and reduce temperatures. As you read on the first page of this lesson, the results of a meteorite crashing into Earth would block sunlight and change climate.

Scientists hypothesize that a meteorite impact might have caused the mass extinction that occurred when dinosaurs became extinct. Evidence for this impact is in a clay layer containing the element iridium in rocks around the world. Iridium is rare in Earth rocks but common in meteorites. No dinosaur fossils have been found in rocks above the iridium layer. A sample of rock containing this layer is shown in **Figure 4.**

 Key Concept Check Describe a possible event that could cause a mass extinction.

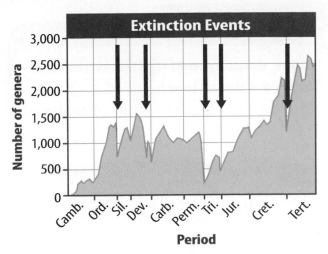

Figure 3 There have been five major mass extinctions in Earth's history. In each one, the number of genera—groups of species—decreased sharply.

✔ **Visual Check** When was Earth's greatest mass-extinction event?

WORD ORIGIN · · · · · · · · · · ·
extinct
from Latin *extinctus,* means "dying out"

Meteorite Impact

Nearly all fossils below the iridium layer in Earth's rocks are different from those above, indicating that a mass extinction occurred.

Figure 4 An iridium-enriched clay layer in Earth's rocks is evidence that a large meteorite crashed into Earth 65 million years ago. A meteorite impact can contribute to a mass extinction event.

Inquiry MiniLab
10 minutes

How does geographic isolation affect evolution?

Have you ever played the phone game? How is this game similar to what happens when populations of organisms are separated?

1. Form two groups.
2. One person in each group should whisper a sentence—provided by your teacher—into the ear of his or her neighbor. Each person in turn will whisper the sentence to his or her neighbor until it returns to the first person.

Analyze and Conclude

1. **Observe** Did the sentence change? Did it change in the same way in each group?

2. **Key Concept** How is this activity similar to organisms that are geographically isolated?

Geography and Evolution

When environments change, some species of organisms are unable to adapt. They become extinct. However, other species do adapt to environmental changes. Evolution is the change in species over time as they adapt to their environments. Sudden, catastrophic changes in the environment can affect evolution. So can the slow movement of Earth's tectonic plates.

Land Bridges When continents collide or when sea level drops, landmasses can join together. *A* **land bridge** *connects two continents that were previously separated.* Over time, organisms move across land bridges and evolve as they adapt to new environments.

Geographic Isolation The movement of tectonic plates or other slow geologic events can cause geographic areas to move apart. When this happens, populations of organisms can become isolated. **Geographic isolation** *is the separation of a population of organisms from the rest of its species due to some physical barrier, such as a mountain range or an ocean.* Separated populations of species evolve in different ways as they adapt to different environments. Even slight differences in environments can affect evolution, as shown in **Figure 5**.

Key Concept Check How can geographic isolation affect evolution?

Geographic Isolation

 **Concepts in Motion** Animation

Figure 5 A population of squirrels was gradually separated as the Grand Canyon developed. Each group adapted to a slightly different environment and evolved in a different way

Kaibab squirrel

Abert's squirrel

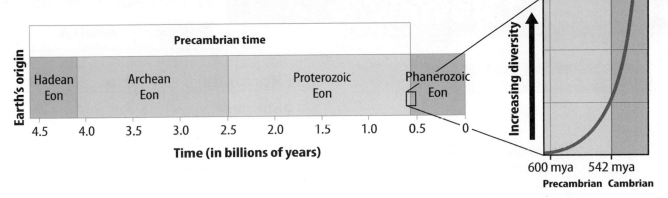

600 mya   542 mya
Precambrian Cambrian

Precambrian Time

Life has been evolving on Earth for billions of years. The oldest fossil evidence of life on Earth is in rocks that are about 3.5 billion years old. These ancient life-forms were simple, unicellular organisms, much like present-day bacteria. The oldest fossils of multicellular organisms are about 600 million years old. These fossils are rare, and early geologists did not know about them. They hypothesized that multicellular life first appeared in the Cambrian (KAM bree un) period, at the beginning of the Phanerozoic eon 542 mya. Time before the Cambrian was called Precambrian time. Scientists have determined that Precambrian time is nearly 90 percent of Earth's history, as shown in **Figure 6.**

Precambrian Life

The rare fossils of multicellular life-forms in Precambrian rocks are from soft-bodied organisms different from organisms on Earth today. A drawing of what they might have looked like is shown in **Figure 7.** Many of these species became extinct at the end of the Precambrian.

Cambrian Explosion

Precambrian life led to a sudden appearance of new types of multicellular life-forms in the Cambrian period. This sudden appearance of new, complex life-forms, indicated on the right in **Figure 6,** is often referred to as the Cambrian explosion. Some Cambrian life-forms, such as trilobites, were the first to have hard body parts. Because of their hard body parts, trilobites were more easily preserved. More evidence of trilobites is in the fossil record. Scientists hypothesize that some of them are distant ancestors of organisms alive today.

Reading Check What is the Cambrian explosion?

Figure 6 Precambrian time is nearly 90 percent of Earth's history. An explosion of life-forms appeared at the beginning of the Phanerozoic eon, during the Cambrian period.

Review
Personal Tutor

Figure 7 Precambrian life-forms lived 600 mya at the bottom of the sea.

Lesson 1 Review

Visual Summary

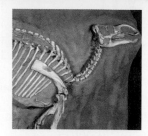

Earth's history is organized into eons, eras, periods, and epochs.

Climate change caused by the results of a meteorite impact could contribute to a mass extinction event.

Slow changes in geography affect evolution.

FOLDABLES

Use your lesson Foldable to review the lesson. Save your Foldable for the project at the end of the chapter.

What do you think NOW?

You first read the statements below at the beginning of the chapter.

1. All geologic eras are the same length of time.

2. Meteorite impacts cause all extinction events.

Did you change your mind about whether you agree or disagree with the statements? Rewrite any false statements to make them true.

Use Vocabulary

1 **Distinguish** between an eon and an era.

2 A(n) _____ might form when continents move close together.

3 A(n) _____ might occur if an environment changes suddenly.

Understand Key Concepts

4 Which could contribute to a mass-extinction event?
   A. an earthquake
   B. a hot summer
   C. a hurricane
   D. a volcanic eruption

5 **Explain** how geographic isolation can affect evolution.

6 **Distinguish** between a calendar and the geologic time scale.

Interpret Graphics

7 **Explain** what the graph below represents. What happened at this time in Earth's past?

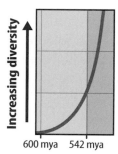

8 **Organize Information** Copy and fill in the graphic organizer below to show units of the geologic time scale from longest to shortest.

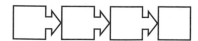

Critical Thinking

9 **Suggest** how humans might contribute to a mass extinction event.

10 **Propose** why Precambrian rocks contain few fossils.

How has life changed over time?

Fossil evidence indicates that there have been wide fluctuations in the types, or diversity, of organisms that have lived on Earth over geologic time.

Learn It

Line graphs compare two variables and show how one variable changes in response to another variable. Line graphs are particularly useful in presenting data that change over time. The first line graph below shows how the diversity of genera has changed over time. The second graph shows how extinction rates, presented as percentages of genera, have changed over time. **Interpret data** in these graphs to learn how they relate to each another.

Try It

1. Carefully study each graph. Note that time, the independent variable, is plotted on the x-axis of each graph. The dependent variable of each graph—the diversity, or number of genera, in one graph and the extinction rate in the other graph—are plotted on the y-axes.

2. Use the graphs to answer questions 3–7.

Apply It

3. According to the graph on the left, at what time in Earth's past was diversity the lowest? At what time was diversity the highest?

4. Approximately what percentage of genera became extinct 250 million years ago?

5. Approximately when did each of Earth's major mass extinctions take place?

6. What is the relationship between diversity and extinction rate?

7. **Key Concept** How have mass extinctions helped scientists develop the geologic time scale?

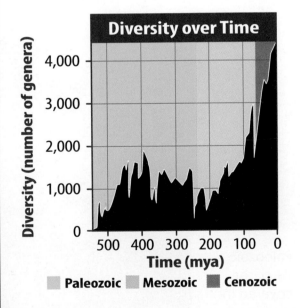

Diversity over Time

Diversity (number of genera): 4,000 / 3,000 / 2,000 / 1,000 / 0

Time (mya): 500 400 300 200 100 0

■ Paleozoic ■ Mesozoic ■ Cenozoic

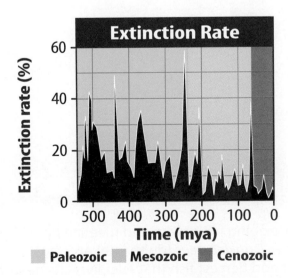

Extinction Rate

Extinction rate (%): 60 / 40 / 20 / 0

Time (mya): 500 400 300 200 100 0

■ Paleozoic ■ Mesozoic ■ Cenozoic

Lesson 2

Reading Guide

Key Concepts 🔑
ESSENTIAL QUESTIONS

- What major geologic events occurred during the Paleozoic era?
- What does fossil evidence reveal about the Paleozoic era?

Vocabulary

Paleozoic era p. 371
Mesozoic era p. 371
Cenozoic era p. 371
inland sea p. 372
coal swamp p. 374
supercontinent p. 375

g Multilingual eGlossary

The Paleozoic Era

Inquiry What animal was this?

Imagine going for a swim and meeting up with this Paleozoic monster. *Dunkleosteus* (duhn kuhl AHS tee us) was one of the largest and fiercest fish that ever lived. Its head was covered in bony armor 5 cm thick—even its eyes had bony armor. It had razor-sharp teethlike plates that bit with a force like that of present-day alligators.

inquiry Launch Lab

20 minutes

What can you learn about your ancestors?

Scientists use fossils and rocks to learn about Earth's history. What could you use to research your past?

1. Write as many facts as you can about one of your grandparents or other older adult family members or friends.

2. What items, such as photos, do you have that can help you?

Think About This

1. If you wanted to know about a great-great-great grandparent, what clues do you think you could find?

2. How does knowledge about past generations in your family benefit you today?

3. 🔑 **Key Concept** How do you think learning about distant relatives is like studying Earth's past?

Early Paleozoic

In many families, three generations—grandparents, parents, and children—live closely together. You could call them the old generation, the middle generation, and the young generation. These generations are much like the three eras of the Phanerozoic eon. *The* **Paleozoic** (pay lee uh ZOH ihk) **era** *is the oldest era of the Phanerozoic eon. The* **Mesozoic** (mez uh ZOH ihk) **era** *is the middle era of the Phanerozoic eon. The* **Cenozoic** (sen uh ZOH ihk) **era** *is the youngest era of the Phanerozoic eon.*

As shown in **Figure 8,** the Paleozoic era lasted for more than half the Phanerozoic eon. Because it was so long, it is often divided into three parts: early, middle, and late. The Cambrian and Ordovician periods make up the Early Paleozoic.

The Age of Invertebrates

The organisms from the Cambrian explosion were invertebrates (ihn VUR tuh brayts) that lived only in the oceans. Invertebrates are animals without backbones. So many kinds of invertebrates lived in Early Paleozoic oceans that this time is often called the age of invertebrates.

WORD ORIGIN · · · · · · · · · · · · · · · ·

Paleozoic
from Greek *palai,* means "ancient"; and Greek *zoe,* means "life"

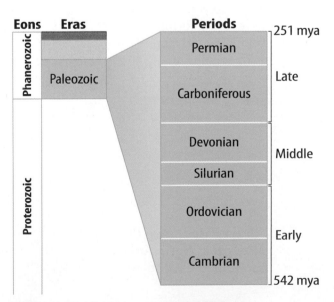

Figure 8 The Paleozoic era lasted for 291 million years. It is divided into six periods.

Cambrian Period
542 – 488
million years ago

Ordovician Period
488 – 444
million years ago

Silurian Period
444 – 416
million years ago

Figure 9 Earth's continents and life-forms changed dramatically during the Paleozoic era.

Visual Check In what period did life first appear on land?

Geology of the Early Paleozoic

If you could have visited Earth during the Early Paleozoic, it would have seemed unfamiliar to you. As shown in **Figure 9,** there was no life on land. All life was in the oceans. The shapes and locations of Earth's continents also would have been unfamiliar, as shown in **Figure 10.** Notice that the landmass that would become North America was on the equator.

Earth's climate was warm during the Early Paleozoic. Rising seas flooded the continents and formed many shallow inland seas. *An **inland sea** is a body of water formed when ocean water floods continents.* Most of North America was covered by an inland sea.

Reading Check How do inland seas form?

FOLDABLES

Make a horizontal, three-tab book. Label it as shown. Use your book to record information about changes during the Paleozoic Era.

Early Paleozoic | Middle Paleozoic | Late Paleozoic

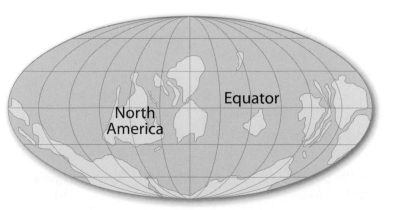

Figure 10 During the Early Paleozoic, North America straddled the equator.

Devonian Period
416 – 359
million years ago

Carboniferous Period
359 – 299
million years ago

Permian Period
299 – 251
million years ago

Middle Paleozoic

The Early Paleozoic ended with a mass extinction event, but many invertebrates survived. New forms of life lived in huge coral reefs along the edges of the continents. Soon, animals with backbones, called vertebrates, evolved.

The Age of Fishes

Some of the earliest vertebrates were fishes. So many types of fishes lived during the Silurian (suh LOOR ee un) and Devonian (dih VOH nee un) periods that the Middle Paleozoic is often called the age of fishes. Some fishes, such as the *Dunkleosteus* pictured at the beginning of this lesson, were heavily armored. **Figure 11** also shows what a *Dunkleosteus* might have looked like. On land, cockroaches, dragonflies, and other insects evolved. Earth's first plants appeared. They were small and lived in water.

Geology of the Middle Paleozoic

Middle Paleozoic rocks contain evidence of major collisions between moving continents. These collisions created mountain ranges. When several landmasses collided with the eastern coast of North America, the Appalachian (ap uh LAY chun) Mountains began to form. By the end of the Paleozoic era, the Appalachians were probably as high as the Himalayas are today.

 Key Concept Check How did the Appalachian Mountains form?

Figure 11 *Dunkleosteus* was a top Devonian predator.

Late Paleozoic

Like the Early Paleozoic, the Middle Paleozoic ended with a mass extinction event. Many marine invertebrates and some land animals disappeared.

The Age of Amphibians

In the Late Paleozoic, some fishlike organisms spent part of their lives on land. *Tiktaalik* (tihk TAH lihk) was an organism that had lungs and could breathe air. It was one of the earliest amphibians. Amphibians were so common in the Late Paleozoic that this time is known as the age of amphibians.

Ancient amphibian species adapted to land in several ways. As you read, they had lungs and could breathe air. Their skins were thick, which slowed moisture loss. Their strong limbs enabled them to move around on land. However, all amphibians, even those living today, must return to the water to mate and lay eggs.

Reptile species evolved toward the end of the Paleozoic era. Reptiles were the first animals that did not require water for reproduction. Reptile eggs have tough, leathery shells that protect them from drying out.

 Key Concept Check How did different species adapt to land?

Coal Swamps

During the Late Paleozoic, dense, tropical forests grew in swamps along shallow inland seas. When trees and other plants died, they sank into the swamps, such as the one illustrated in **Figure 12.** A **coal swamp** *is an oxygen-poor environment where, over time, plant material changes into coal.* The coal swamps of the Carboniferous (car buhn IF er us) and Permian periods eventually became major sources of coal that we use today.

Figure 12 Plants buried in ancient coal swamps became coal.

Formation of Pangaea

Geologic evidence indicates that many continental collisions occurred during the Late Paleozoic. As continents moved closer together, new mountain ranges formed. By the end of the Paleozoic era, Earth's continents had formed a giant supercontinent—Pangaea. A **supercontinent** *is an ancient landmass which separated into present-day continents*. Pangaea formed close to Earth's equator, as shown in **Figure 13.** As Pangaea formed, coal swamps dried up and Earth's climate became cooler and drier.

The Permian Mass Extinction

The largest mass extinction in Earth's history occurred at the end of the Paleozoic era. Fossil evidence indicates that 95 percent of marine life-forms and 70 percent of all life on land became extinct. This extinction event is called the Permian mass extinction.

 Key Concept Check What does fossil evidence reveal about the end of the Paleozoic era?

Scientists debate what caused this mass extinction. The formation of Pangaea likely decreased the amount of space where marine organisms could live. It would have contributed to changes in ocean currents, making the center of Pangaea drier. But Pangaea formed over many millions of years. The extinction event occurred more suddenly.

Some scientists hypothesize that a large meteorite impact caused drastic climate change. Others propose that massive volcanic eruptions changed the global climate. Both a meteorite impact and large-scale eruptions would have ejected ash and rock into the atmosphere, blocking out sunlight, reducing temperatures, and causing a collapse of food webs.

Whatever caused it, Earth had fewer species after the Permian mass extinction. Only species that could adapt to the changes survived.

Pangaea

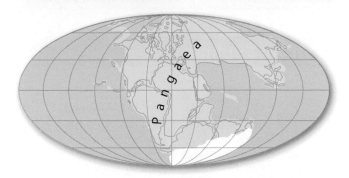

Figure 13 The supercontinent Pangaea formed at the end of the Paleozoic era.

Concepts in Motion Animation

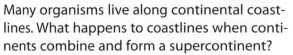

MiniLab 20 minutes

What would happen if a supercontinent formed?

Many organisms live along continental coastlines. What happens to coastlines when continents combine and form a supercontinent?

1. Read and complete a lab safety form.
2. Form a stick of **modeling clay** into a flat pancake shape. Form three pancake shapes from an identical stick of clay. Make all four shapes equal thicknesses.
3. With a **flexible tape measure,** measure the perimeter of each shape.

Analyze and Conclude

1. **Compare** Is the perimeter of the larger shape more or less than the combined perimeters of the three smaller shapes?

2. **Key Concept** How might the formation of Pangaea have affected life on Earth?

Lesson 2 Review

Visual Summary

Life slowly moved to land during the Paleozoic era as amphibians and reptiles evolved.

In the Late Paleozoic, massive coal swamps formed along inland seas.

At the end of the Paleozoic era, a mass extinction event coincided with the final stages of the formation of Pangaea.

FOLDABLES®

Use your lesson Foldable to review the lesson. Save your Foldable for the project at the end of the chapter.

What do you think NOW?

You first read the statements below at the beginning of the chapter.

3. North America was once on the equator.

4. All of Earth's continents were part of a huge supercontinent 250 million years ago.

Did you change your mind about whether you agree or disagree with the statements? Rewrite any false statements to make them true.

Use Vocabulary

1. **Distinguish** between the Paleozoic era and the Mesozoic era.

2. When ocean water covers part of a continent, a(n) _____ forms.

3. **Use the term** *supercontinent* in a complete sentence.

Understand Key Concepts

4. Which was true of North America during the Early Paleozoic?
   A. It had many glaciers.
   B. It was at the equator.
   C. It was part of a supercontinent.
   D. It was populated by reptiles.

5. **Compare** ancient amphibians and reptiles and explain how each group adapted to live on land.

6. **Draw** a cartoon that shows how the Appalachian Mountains formed.

Interpret Graphics

7. **Organize** A time line of the Paleozoic era is pictured below. Copy the time line and fill in the missing periods.

Paleozoic				
	Ordovician	Silurian	Devonian	Carboniferous	

8. **Sequence** Copy and fill in the graphic organizer below. Start with Precambrian time, then list the eras in order.

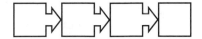

Critical Thinking

9. **Consider** What if 100 percent of organisms had become extinct at the end of the Paleozoic era?

10. **Evaluate** the possible effects of climate change on present-day organisms.

When did coal form?

Coal is fossilized plant material. When swamp plants die, they become covered by oxygen-poor water and change to peat. Over time, high temperatures and pressure from sediments transform the peat into coal. When did the plants live that formed the coal we use today?

Learn It

A bar graph can display the same type of information as a line graph. However, instead of data points and a line that connects them, a bar graph uses rectangular bars to show how values compare. **Interpret the data** below to learn when most coal formed.

Try It

1 Carefully study the bar graph. Notice that time is plotted on the *x*-axis (as geologic periods), and that coal deposits (as tons accumulated per year) are plotted on the *y*-axis.

2 Use the graph and what you know about coal formation to answer the following questions.

Apply It

3 Which coal deposits are oldest? Which are youngest?

4 During which geologic period did most of the coal form?

5 Approximately how much coal accumulated during the Paleozoic era? The Mesozoic era?

6 Why are there no data on the graph for the Cambrian, Ordovician, and Silurian periods of geologic time?

7 🗝 **Key Concept** What does fossil evidence reveal about the Paleozoic era?

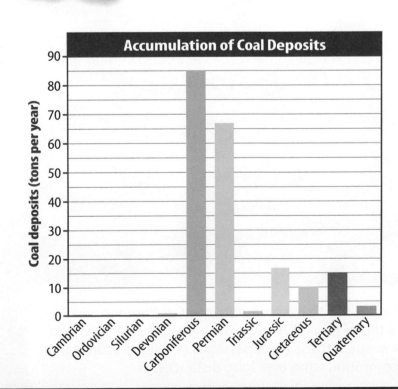

Accumulation of Coal Deposits

Coal deposits (tons per year)

Cambrian, Ordovician, Silurian, Devonian, Carboniferous, Permian, Triassic, Jurassic, Cretaceous, Tertiary, Quaternary

Reading Guide

Key Concepts 🔑
ESSENTIAL QUESTIONS

- What major geologic events occurred during the Mesozoic era?

- What does fossil evidence reveal about the Mesozoic era?

Vocabulary

dinosaur p. 382

plesiosaur p. 383

pterosaur p. 383

g **Multilingual eGlossary**

The Mesozoic Era

Inquiry Mesozoic Thunder?

Can you imagine the sounds this dinosaur made? *Corythosaurus* had a tall, bony crest on top of its skull. Long nasal passages extended into the crest. Scientists suspect these nasal passages amplified sounds that could be used for communicating over long distances.

How diverse were dinosaurs?

How many different dinosaurs were there?

1. Read and complete a lab safety form.

2. Your teacher will give you an **index card** listing a species name of a dinosaur, the dinosaur's dimensions, and the time when it lived.

3. Draw a picture of what you imagine your dinosaur looked like. Before you begin, decide with your classmates what common scale you should use.

4. **Tape** your dinosaur drawing to the Mesozoic time line your teacher provides.

Think About This

1. What was the biggest dinosaur? The smallest? Can you see any trends in size on the time line?

2. Did all the dinosaurs live at the same time?

3. ⚷ **Key Concept** Dinosaurs were numerous and diverse. Do you think any dinosaurs could swim or fly?

Geology of the Mesozoic Era

When people imagine what Earth looked like millions of years ago, they often picture a scene with dinosaurs, such as the *Corythosaurus* shown on the opposite page. Dinosaurs lived during the Mesozoic era. The Mesozoic era lasted from 251 mya to 65.5 mya. As shown in **Figure 14,** it is divided into three periods: the Triassic (tri A sihk), the Jurassic (joo RA sihk), and the Cretaceous (krih TAY shus).

Breakup of Pangaea

Recall that the supercontinent Pangaea formed at the end of the Paleozoic era. The breakup of Pangaea was the dominant geologic event of the Mesozoic era. Pangaea began to break apart in the Late Triassic. Eventually, Pangaea split into two separate landmasses—Gondwanaland (gahn DWAH nuh land) and Laurasia (la RAY shzah). Gondwanaland was the southern continent. It included the future continents of Africa, Antarctica, Australia, and South America. Laurasia, the northern continent, included the future continents of North America, Europe, and Asia.

FOLDABLES

Make a shutter-fold book from a vertical sheet of paper. Label it as shown. Use it to record information about changes during the Mesozoic era.

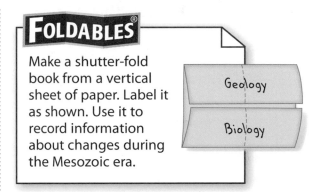

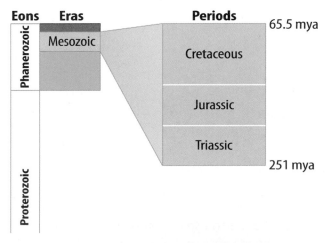

Figure 14 The Mesozoic era was the middle era of the Phanerozoic eon. It lasted for 185.5 million years.

Triassic Period
251.0 – 201.6 million years ago

Figure 15 Dinosaurs dominated the Mesozoic era, but many other species also lived during this time in Earth's history.

Return of Shallow Seas

The type of species represented in **Figure 15** adapted to an environment of lush tropical forests and warm ocean waters. That is because the climate of the Mesozoic era was warmer than the climate of the Paleozoic era. It was so warm that, for most of the era, there were no ice caps, even at the poles. With no glaciers, the oceans had more water. Some of this water flowed onto the continents as Pangaea split apart. This created narrow channels that grew larger as the continents moved apart. Eventually, the channels became oceans. The Atlantic Ocean began to form at this time.

🔑 **Key Concept Check** When did the Atlantic Ocean begin to form?

Sea level rose during most of the Mesozoic era, as shown in **Figure 16.** Toward the end of the era, sea level was so high that inland seas covered much of Earth's continents. This provided environments for the evolution of new organisms.

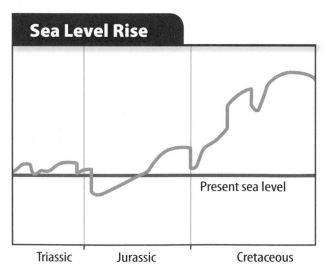

Sea Level Rise

Present sea level

Triassic Jurassic Cretaceous

Figure 16 Earth's sea level rose during the Mesozoic era.

 Visual Check In which period was sea level at its highest?

Jurassic Period		Cretaceous Period
201.6 – 145.5 million years ago		145.5 – 65.5 million years ago

Mesozoic North America

Along North America's eastern coast and the Gulf of Mexico, sea level rose and receded over millions of years. As this happened, sea-water **evaporated,** leaving massive salt deposits behind. Some of these salt deposits are sources of salt today. Other salt deposits later became traps for oil. Today, salt traps in the Gulf of Mexico are an important source of oil.

Throughout the Mesozoic era, the North American continent moved slowly and steadily westward. Its western edge collided with several small landmasses carried on an ancient oceanic plate. As this plate subducted beneath the North American continent, the crust buckled inland, slowly pushing up the Rocky Mountains, shown on the map in **Figure 17.** In the dry southwest, windblown sand formed huge dunes. In the middle of the continent, a warm inland sea formed.

Key Concept Check How did the Rocky Mountains form?

Figure 17 The Rocky Mountains began forming during the Mesozoic era. By the end of the era, an inland sea covered much of the central part of North America.

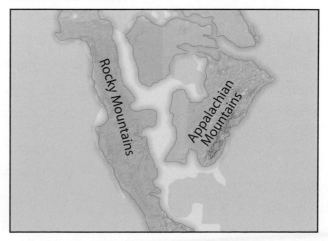

Can you run like a reptile?

Unlike the limbs of crocodiles and other modern reptiles, dinosaur limbs were positioned directly under their bodies. What did this mean?

1. Pick a partner. One of you—the dinosaur—will run on all fours with arms held straight directly below the shoulders. The other—the crocodile—will run with arms bent and positioned out from the body.

2. Race each other, then reverse positions.

Analyze and Conclude

1. **Compare** Which could move faster—the dinosaur or the crocodile?

2. **Infer** Which posture do you think could support more weight?

3. 🔑 **Key Concept** How might their posture have enabled dinosaurs to become so successful? How might it have helped them become so large?

Figure 18 Fossils provide evidence that the hip structure of a dinosaur enabled it to walk upright.

Mesozoic Life

The species of organisms that survived the Permian mass extinction event lived in a world with few species. Vast amounts of unoccupied space were open for organisms to inhabit. New types of cone-bearing trees, such as pines and cycads, began to appear. Toward the end of the era, the first flowering plants evolved. Dominant among vertebrates living on land were the dinosaurs. Hundreds of species of many sizes existed.

Dinosaurs

Though dinosaurs have long been considered reptiles, scientists today actively debate dinosaur classification. Dinosaurs share a common ancestor with present-day reptiles, such as crocodiles. However, dinosaurs differ from present-day reptiles in their unique hip structure, as shown in **Figure 18**. **Dinosaurs** *were dominant Mesozoic land vertebrates that walked with legs positioned directly below their hips.* This meant that many walked upright. In contrast, the legs of a crocodile stick out sideways from its body. It appears to drag itself along the ground.

Scientists hypothesize that some dinosaurs are more closely related to present-day birds than they are to present-day reptiles. Dinosaur fossils with evidence of feathery exteriors have been found. For example, *Archaeopteryx* (ar kee AHP tuh rihks), a small bird the size of a pigeon, had wings and feathers but also claws and teeth. Many scientists suggest it was an ancestor to birds.

Dinosaur Posture 🔑

Erect posture

Sprawling posture

Sprawling posture

Erect posture

Other Mesozoic Vertebrates

Dinosaurs dominated land. But, fossils indicate that other large vertebrates swam in the seas and flew in the air, as shown in **Figure 19**. **Plesiosaurs** (PLY zee oh sorz) *were Mesozoic marine reptiles with small heads, long necks, and flippers.* Through much of the Mesozoic, these reptiles dominated the oceans. Some were as long as 14 m.

Other Mesozoic reptiles could fly. **Pterosaurs** (TER oh sorz) *were Mesozoic flying reptiles with large, batlike wings.* One of the largest pterosaurs, the *Quetzalcoatlus* (kwetz oh koh AHT lus), had a wingspread of nearly 12 m. Though they could fly, pterosaurs were not birds. As you have read, birds are more closely related to dinosaurs.

 Key Concept Check How could you distinguish fossils of plesiosaurs and pterosaurs from fossils of dinosaurs?

Appearance of Mammals

Dinosaurs and reptiles dominated the Mesozoic era, but another kind of animal also lived during this time—mammals. Mammals evolved early in the Mesozoic and remained small in size throughout the era. Few were larger than present-day cats.

Cretaceous Extinction Event

The Mesozoic era ended 65.5 mya with a mass extinction called the Cretaceous extinction event. You read in Lesson 1 that scientists propose a large meteorite impact contributed to this extinction. This crash would have produced enough dust to block sunlight for a long time. There is evidence that volcanic eruptions also occurred at the same time. These eruptions would have added more dust to the atmosphere. Without light, plants died. Without plants, animals died. Dinosaur species and other large Mesozoic vertebrate species could not adapt to the changes. They became extinct.

Figure 19 Not all large Mesozoic vertebrates were dinosaurs.

Visual Check How did the limbs of these reptiles compare to the limbs of dinosaurs?

WORD ORIGIN

pterosaur
from Greek *pteron*, means "wing"; and *sauros*, means "lizard"

Lesson 3 Review

Visual Summary

As Pangaea broke up, the continents began to move into their present-day positions.

Rocky Mountains

The Mesozoic climate was warm and sea level was high.

Dinosaurs were not the only large vertebrates that lived during the Mesozoic era.

FOLDABLES®

Use your lesson Foldable to review the lesson. Save your Foldable for the project at the end of the chapter.

What do you think NOW?

You first read the statements below at the beginning of the chapter.

5. All large Mesozoic vertebrates were dinosaurs.

6. Dinosaurs disappeared in a large mass extinction event.

Did you change your mind about whether you agree or disagree with the statements? Rewrite any false statements to make them true.

Use Vocabulary

1 A(n) _____ was a marine Mesozoic reptile.

2 A(n) _____ was a Mesozoic reptile that could fly.

Understand Key Concepts

3 Which major event happened during the Mesozoic era?
   A. Humans evolved.
   B. Life moved onto land.
   C. The Appalachian Mountains formed.
   D. The Atlantic Ocean formed.

4 **Compare** the sizes of reptiles and mammals during the Mesozoic era.

5 **Explain** how the Rocky Mountains formed.

Interpret Graphics

6 **Identify** Which type of vertebrate does each skeletal figure below represent?

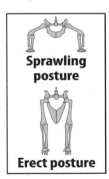

Sprawling posture

Erect posture

7 **Sequence** Copy and fill in the graphic organizer below to list the periods of the Mesozoic era in order.

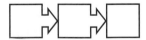

Critical Thinking

8 **Infer** how Earth might be different if there had been no extinction event at the end of the Mesozoic era.

9 **Propose** how the breakup of Pangaea might have affected evolution.

AMERICAN
MUSEUM OF
NATURAL
HISTORY

CAREERS
in SCIENCE

Digging Up a Surprise

A fossil discovery in China reveals some unexpected clues about early mammals.

The Mesozoic era, 251 to 65.5 million years ago, was the age of the dinosaurs. Many species of dinosaurs roamed Earth, from the ferocious tyrannosaurs to the giant, long-necked brachiosaurs. What other animals lived among the dinosaurs? For years, paleontologists assumed that the only mammals that lived at that time were no bigger than mice. They were no match for the dinosaurs.

Recent fossil discoveries revealed new information about these early mammals. Jin Meng is a paleontologist at the American Museum of Natural History in New York City. In northern China, Meng and other paleontologists discovered fossils of animals that probably died in volcanic eruptions 130 million years ago. Among these fossils were the remains of a mammal over 1 foot long—about the size of a small dog. A representation of the mammal, *Repenomamus robustus* (reh peh noh MA muhs • roh BUS tus), is shown to the right.

▲ Paleontologists studying a fossil of the mammal *Repenomamus robustus* found tiny *Psittacosaurus* bones in its stomach.

This fossil would reveal an even bigger surprise. When examined under microscopes in the lab, scientists discovered small bones in the fossil's rib cage where its stomach had been. The bones were the tiny limbs, fingers, and teeth of a young plant-eating dinosaur. The mammal's last meal had been a young dinosaur!

This was an exciting discovery. Meng and his team learned that early mammals were larger than they thought and were meat eaters, too. Those tiny bones proved to be a huge find. Paleontologists now have a new picture of how animals interacted during the age of dinosaurs.

This is a representation of a young *Psittacosaurus*—only 12 cm long. ▶

It's Your Turn

DIAGRAM With a group, research the plants and the animals that lived in the same environment as *Repenomamus*. Create a drawing showing the relationships among the organisms. Compare your drawing to those of other groups.

Reading Guide

Key Concepts 🔑
ESSENTIAL QUESTIONS

- What major geologic events occurred during the Cenozoic era?

- What does fossil evidence reveal about the Cenozoic era?

Vocabulary

Holocene epoch p. 387

Pleistocene epoch p. 389

ice age p. 389

glacial groove p. 389

mega-mammal p. 390

g **Multilingual eGlossary**

The Cenozoic Era

Inquiry Is this animal alive?

No, this is a statue in a Los Angeles, California, pond that has been oozing tar for thousands of years. It shows how a mammoth might have become stuck in a tar pit. Mammoths lived at the same time as early humans. What do you think it was like to live alongside these animals?

What evidence do you have that you went to kindergarten?

Rocks and fossils provide evidence about Earth's past. The more recent the era, the more evidence exists. Is this true for you, too?

1 Make a list of items you have, such as a diploma, that could provide evidence about what you did and what you learned in kindergarten.

2 Make another list of items that could provide evidence about your school experience during the past year.

Think About This

1. Which list is longer? Why?

2. 🔑 **Key Concept** How do you think the items on your lists are like evidence from the first and last eras of the Phanerozoic eon?

Geology of the Cenozoic Era

Have you ever experienced a severe storm? What did your neighborhood look like afterward? Piles of snow, rushing water, or broken trees might have made your neighborhood seem like a different place. In a similar way, the landscapes and organisms of the Paleozoic and Mesozoic eras might have been strange and unfamiliar to you. Though some unusual animals lived during the Cenozoic era, this era is more familiar. People know more about the Cenozoic era than they know about any other era because we live in the Cenozoic era. Its fossils and its rock record are better preserved.

As shown in **Figure 20,** the Cenozoic era spans the time from the end of the Cretaceous period, 65.5 mya, to present day. Geologists divide it into two periods—the Tertiary (TUR shee ayr ee) period and the Quaternary (KWAH tur nayr ee) period. These periods are further subdivided into epochs. *The most recent epoch, the* **Holocene** *(HOH luh seen)* **epoch,** *began 10,000 years ago.* You live in the Holocene epoch.

FOLDABLES®

Make a shutter-fold book from a vertical sheet of paper. Label it as shown. Use it to record information about changes during the Cenozoic era.

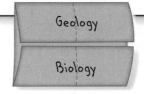

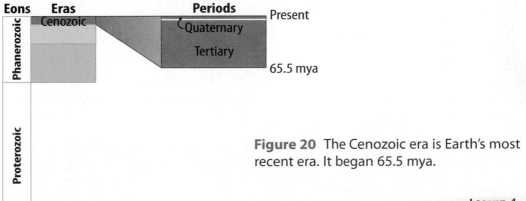

Figure 20 The Cenozoic era is Earth's most recent era. It began 65.5 mya.

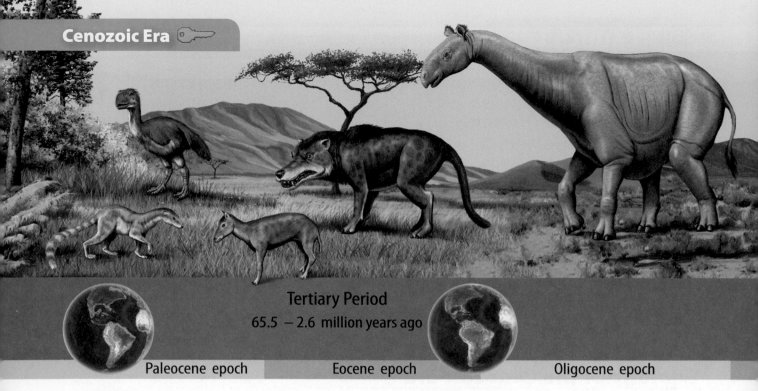

Tertiary Period
65.5 − 2.6 million years ago

Paleocene epoch   Eocene epoch   Oligocene epoch

Figure 21 Mammals dominated the landscapes of the Cenozoic era.

WORD ORIGIN
Cenozoic
From Greek *kainos*, means "new"; and *zoic*, means "life"

 Review

Math Skills
Math Practice
Personal Tutor

Use Percentages
The Cenozoic era began 65.5 mya. What percentage of the Cenozoic era is taken up by the Quaternary period, which began 2.6 mya? To calculate the percentage of a part to the whole, perform the following steps:

a. Express the problem as a fraction.

$$\frac{2.6 \text{ mya}}{65.5 \text{ mya}}$$

b. Convert the fraction to a decimal. 2.6 mya divided by 65.5 mya = 0.040

c. Multiply by 100 and add %.
0.040 × 100 = 4.0%

Practice
What percent of the Cenozoic era is represented by the Tertiary period, which lasted from 65.5 mya to 2.6 mya? [Hint: Subtract to find the length of the Tertiary period.]

Cenozoic Mountain Building
As shown in the globes in **Figure 21,** Earth's continents continued to move apart during the Cenozoic era, and the Atlantic Ocean continued to widen. As the continents moved, some landmasses collided. Early in the Tertiary period, India crashed into Asia. This collision began to push up the Himalayas—the highest mountains on Earth today. At about the same time, Africa began to push into Europe, forming the Alps. These mountains continue to get higher today.

In North America, the western coast continued to push against the seafloor next to it, and the Rocky Mountains continued to grow in height. New mountain ranges—the Cascades and the Sierra Nevadas—began to form along the western coast. On the eastern coast, there was little tectonic activity. The Appalachian Mountains, which formed during the Paleozoic era, continue to erode today.

Reading Check Why are the Appalachian Mountains relatively small today?

| Miocene epoch | Pliocene epoch | Pleistocene epoch | Holocene epoch |

Quaternary Period
2.6 million years — present

Pleistocene Ice Age

Like the Mesozoic era, the early part of the Cenozoic era was warm. In the middle of the Tertiary period, the climate began to cool. By the Pliocene (PLY oh seen) epoch, ice covered the poles as well as many mountaintops. It was even colder during the next epoch—the Pleistocene (PLY stoh seen).

The **Pleistocene epoch** *was the first epoch of the Quaternary period.* During this time, glaciers advanced and retreated many times. They covered as much as 30 percent of Earth's land surface. *An* **ice age** *is a time when a large proportion of Earth's surface is covered by glaciers.* Sometimes, rocks carried by glaciers created deep gouges or grooves, as shown in **Figure 22.** **Glacial grooves** *are grooves made by rocks carried in glaciers.*

The glaciers contained huge amounts of water. This water originated in the oceans. With so much water in glaciers, sea level dropped. As sea level dropped, inland seas drained away, exposing dry land. When sea level was at its lowest, the Florida peninsula was about twice as wide as it is today.

Pleistocene Ice Age 🔑

Figure 22 Glacial grooves in Ohio are evidence that glaciers extended far into North America during the Pleistocene ice age.

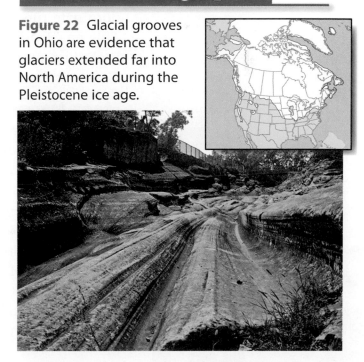

✓ **Visual Check** Approximately what percentage of the United States was covered with ice?

Figure 23 These mega-mammals lived at different times during the Cenozoic era. They are all extinct today. The human is included for reference.

Cenozoic Life—The Age of Mammals

The mass extinction event at the end of the Mesozoic era meant that there was more space for each surviving species. Flowering plants, including grasses, evolved and began to dominate the land. These plants provided new food sources. This enabled the evolution of many types of animal species, including mammals. Mammals were so successful that the Cenozoic era is sometimes called the age of mammals.

Mega-Mammals

Recall that mammals were small during the Mesozoic era. Many new types of mammals appeared during the Cenozoic era. Some were very large, such as those shown in **Figure 23.** *The large mammals of the Cenozoic era are called* **mega-mammals.** Some of the largest lived during the Oligocene and Miocene periods, from 34 mya to 5 mya. Others, such as woolly mammoths, giant sloths, and saber-toothed cats, lived during the cool climate of the Pliocene and Pleistocene periods, from 5 mya to 10,000 years ago. Many fossils of these animals have been discovered. The saber-toothed cat skull in **Figure 24** was discovered in the Los Angeles tar pits pictured at the beginning of this lesson. A few mummified mammoth bodies also have been discovered preserved for thousands of years in glacial ice.

Figure 24 🔑 The saber-toothed cat was a fierce Pleistocene predator.

🔑 **Key Concept Check** How do scientists know that mega-mammals lived during the Cenozoic era?

Isolated Continents and Land Bridges

The mammals depicted in **Figure 23** lived in North America, South America, Europe, and Asia. Different mammal species evolved in Australia. This is mostly because of the movement of Earth's tectonic plates. You read earlier that land bridges can connect continents that were once separated. You also read that when continents are separated, species that once lived together can become geographically isolated.

Most of the mammals that live in Australia today are marsupials (mar SOO pee ulz). These mammals, like kangaroos, carry their young in pouches. Some scientists suggest that marsupials did not evolve in Australia. Instead, they **hypothesize** that marsupial ancestors migrated to Australia from South America when South America and Australia were connected to Antarctica by land bridges, as shown in **Figure 25**. After ancestral marsupials arrived in Australia, Australia moved away from Antarctica, and water covered the land bridges between South America, Antarctica, and Australia. Over time, the ancestral marsupials evolved into the types of marsupials that live in Australia today.

 Reading Check What major geologic events affected the evolution of marsupials in Australia?

ACADEMIC VOCABULARY

hypothesize
(verb) To make an assumption about something that is not positively known

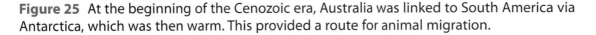

Land Bridges

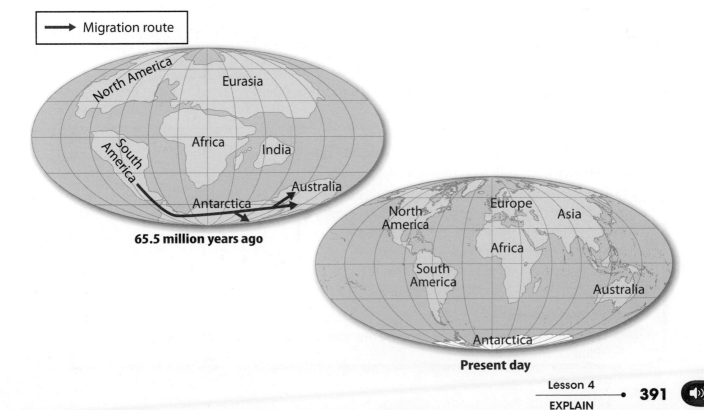

Figure 25 At the beginning of the Cenozoic era, Australia was linked to South America via Antarctica, which was then warm. This provided a route for animal migration.

→ Migration route

65.5 million years ago

Present day

Rise of Humans

The oldest fossil remains of human ancestors have been found in Africa, where scientists think humans first evolved. These fossils are nearly 6 million years old. A skeleton of a 3.2-million-year-old human ancestor is shown in **Figure 26.**

Modern humans—called *Homo sapiens*—didn't evolve until the Pleistocene epoch. Early *Homo sapiens* migrated to Europe, Asia, and eventually North America. Early humans likely migrated to North America from Asia using a land bridge that connected the continents during the Pleistocene ice age. This land bridge is now covered with water.

Pleistocene Extinctions

Climate changed at the close of the Pleistocene epoch 10,000 years ago. The Holocene epoch was warmer and drier. Forests replaced grasses. The mega-mammals that lived during the Pleistocene became extinct. Some scientists suggest that mega-mammal species could not adapt fast enough to survive the environmental changes.

Key Concept Check How did climate change at the end of the Pleistocene epoch?

Future Changes

There is evidence that present-day Earth is undergoing a global-warming climate change. Many scientists suggest that humans have contributed to this change because of their use of coal, oil, and other fossil fuels over the past few centuries.

Figure 26 *Lucy* is the name scientists have given this 3.2-million-year-old human ancestor.

Inquiry MiniLab   20 minutes

What happened to the Bering land bridge?

Pleistocene animals and humans likely crossed into North America from Asia using the Bering land bridge. Why did this bridge disappear?

1. Read and complete a lab safety form
2. Form two pieces of **modeling clay** into continents, each with a continental shelf.
3. Place the clay models into a **watertight container** with the continental shelves touching. Add water, leaving the continental shelves exposed. Place a dozen or more **ice cubes** on the continents.
4. During your next science class, observe the container and record your observations.

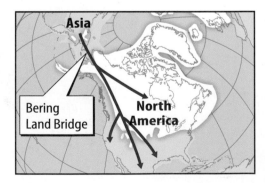

Analyze and Conclude

Key Concept How does your model represent what happened at the end of the Pleistocene epoch?

Lesson 4 Review

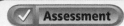

Visual Summary

The mega-mammals that lived during most of the Cenozoic era are extinct.

Glaciers extended well into North America during the Pleistocene ice age.

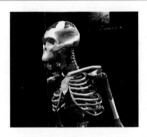

Lucy is a 3.2-million-year-old human ancestor.

FOLDABLES®

Use your lesson Foldable to review the lesson. Save your Foldable for the project at the end of the chapter.

What do you think **NOW?**

You first read the statements below at the beginning of the chapter.

7. Mammals evolved after dinosaurs became extinct.

8. Ice covered nearly one-third of Earth's land surface 10,000 years ago.

Did you change your mind about whether you agree or disagree with the statements? Rewrite any false statements to make them true.

Use Vocabulary

1 Gouges made by ice sheets are _____.

2 You live in the _____ epoch.

Understand Key Concepts

3 Which organism lived during the Cenozoic era?
- **A.** *Brachiosaurus*
- **B.** *Dunkleosteus*
- **C.** saber-toothed cats
- **D.** trilobites

4 **Classify** Which terms are associated with the Cenozoic era: *Homo sapiens*, mammoth, dinosaur, grass?

Interpret Graphics

5 **Determine** The map below shows coastlines of the southeastern U.S. at three times during the Cenozoic era. Which choice represents the coastline at the height of the Pleistocene ice age?

Choice A
Choice B
Choice C

6 **Summarize** Copy and fill in the graphic organizer below to list living mammals that might be considered mega-mammals today.

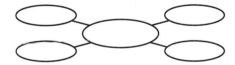

Critical Thinking

7 **Suggest** what might happen if the Australian continent crashed into Asia.

Math Skills ×÷+−

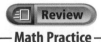 Review
—— Math Practice ——

8 The Cenozoic era began 65.5 mya. The Oligocene and Miocene epochs extended from 34 mya to 5 mya. What percentage of the Cenozoic era is represented by the Oligocene and Miocene epochs?

Modeling Geologic Time

Materials

meterstick

tape measure

poster board

colored markers

colored paper

string

maps

Evidence suggests that Earth formed approximately 4.6 billion years ago. But how long is 4,600,000,000 years? It is difficult to comprehend time that extends so far into the past unless you can relate it to your own experience. In this activity, you will develop a metaphor for geologic time using a scale that is familiar to you. Then, you will create a model to share with your class.

Question

How can you model geologic time using a familiar scale?

Procedure

1 Think of something you are familiar with that can model a long period of time. For example, you might choose the length of a football field or the distance between two U.S. cities on a map—one on the east coast and one on the west coast.

2 Make a model of your metaphor using a metric scale. On your model, display the events listed in the table on the next page. Use the equation below to generate true-to-scale dates in your model.

$$\frac{\text{Known age of past event (years before present)}}{\text{Known age of Earth (years before present)}} = \frac{X \text{ time scale unit location}}{\text{Maximum distance or extent of metaphor}}$$

Example: To find where "first fish" would be placed on your model if you used a meterstick (100 cm), set up your equation as follows:

$$\frac{500,000,000 \text{ years}}{4,600,000,000 \text{ years}} = \frac{X \text{ (location on meterstick)}}{100 \text{ cm}}$$

3 In your Science Journal, keep a record of all the math equations you used. You can use a calculator, but show all equations.

Analyze and Conclude

4 **Calculate** What percentage of geologic time have modern humans occupied? Set up your equation as follows:

$$\frac{100,000}{4,600,000,000} \times 100 = \text{\% of time occupied by } H. \text{ sapiens}$$

5 **Estimate** Where does the Precambrian end on your model? Estimate how much of geologic time falls within the Precambrian.

6 **Evaluate** What other milestone events in Earth's history, other than those listed in the table, could you include on your model?

7 **Appraise** the following sentence as it relates to your life: "Time is relative."

8  **The Big Idea** The Earth events on your model are based mostly on fossil evidence. How are fossils useful in understanding Earth's history? How are they useful in the development of the geologic time scale?

Communicate Your Results

Share your model with the class. Explain why you chose the model you did, and demonstrate how you calculated the scale on your model.

Inquiry Extension

Imagine that you were asked to teach a class of kindergartners about Earth's time. How would you do it? What metaphor would you use? Why?

Some Important Approximate Dates in the History of Earth:

MYA	Event
4,600	Origin of Earth
3,500	Oldest evidence of life
500	First fish
375	Tiktaalik appears
320	First reptiles
250	Permian extinction event
220	Mammals and dinosaurs appear
155	Archaeopteryx appears
145	Atlantic Ocean forms
65	Cretaceous extinction event
6	Human ancestors appear
2	Pleistocene Ice Age begins
0.1	Homo sapiens appear
0.00052	Columbus lands in New World
??	Your birth date

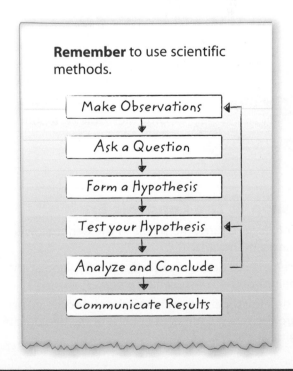

Remember to use scientific methods.

Make Observations → Ask a Question → Form a Hypothesis → Test your Hypothesis → Analyze and Conclude → Communicate Results

Chapter 11 Study Guide

 WebQuest

 THE BIG IDEA The geologic changes that have occurred during the billions of years of Earth's history have strongly affected the evolution of life.

Key Concepts Summary

Lesson 1: Geologic History and the Evolution of Life

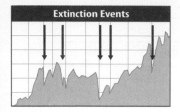

Extinction Events

- Geologists organize Earth's history into **eons, eras, periods,** and **epochs.**
- Life evolves over time as Earth's continents move, forming **land bridges** and causing **geographic isolation.**
- **Mass extinctions** occur if many species of organisms cannot adapt to sudden environmental change.

Lesson 2: The Paleozoic Era

- Life diversified during the **Paleozoic era** as organisms moved from water to land.
- **Coal swamps** formed along **inland seas.** Later, land became drier as the **supercontinent** Pangaea formed.
- The largest mass extinction in Earth's history occurred at the end of the Permian period.

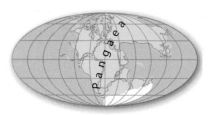

Lesson 3: The Mesozoic Era

- Sea level rose as the climate warmed.
- The Atlantic Ocean and the Rocky Mountains began to form as Pangaea broke apart.
- **Dinosaurs, plesiosaurs, pterosaurs,** and other large Mesozoic vertebrates became extinct at the end of the era.

Lesson 4: The Cenozoic Era

- The large, extinct mammals of the Cenozoic were **mega-mammals.**
- Ice covered nearly one-third of Earth's land at the height of the Pleistocene **ice age.**
- The **Pleistocene epoch** and the **Holocene epoch** are the two most recent epochs of the geologic time scale.

Vocabulary

eon p. 363
era p. 363
period p. 363
epoch p. 363
mass extinction p. 365
land bridge p. 366
geographic isolation p. 366

Paleozoic era p. 371
Mesozoic era p. 371
Cenozoic era p. 371
inland sea p. 372
coal swamp p. 374
supercontinent p. 375

dinosaur p. 382
plesiosaur p. 383
pterosaur p. 383

Holocene epoch p. 387
Pleistocene epoch p. 389
ice age p. 389
glacial groove p. 389
mega-mammal p. 390

FOLDABLES® Chapter Project

Assemble your lesson Foldables as shown to make a Chapter Project. Use the project to review what you have learned in this chapter.

Geologic Time

Mesozoic Era

Paleozoic Era

Cenozoic Era

Use Vocabulary

1. The longest time unit in the geologic time scale is the _____.

2. Eras are subdivided into _____.

3. Many boundaries in the geologic time scale are marked by the occurrence of _____.

4. When glaciers melt, shallow _____ form in the interiors of continents.

5. The _____ was the first era of the Phanerozoic eon.

6. A(n) _____ can form when plants are buried in an oxygen-poor environment.

7. Marine Mesozoic reptiles included _____.

8. Modern humans evolved during the _____.

Link Vocabulary and Key Concepts

Concepts in Motion Interactive Concept Map

Copy this concept map, and then use vocabulary terms from the previous page and other terms from the chapter to complete the concept map.

Geologic Time
is subdivided into (in order)

| 9 | 10 | 11 | 12 |

in which early life-forms, such as

13

appear.

Life was in the

14

Two major ages

| 15 | 16 |

Ends in

17

Life was evolving on

18

Time of the

19

Ends in

20

Age of the

21

Characterized by most recent

22

Understand Key Concepts

1 The trilobite fossil below represents an organism that lived during the Cambrian period.

What distinguished this organism from organisms that lived earlier in time?

A. It had hard parts.
B. It lived on land.
C. It was a reptile.
D. It was multicellular.

2 What are the many divisions in the geologic time scale based on?

A. changes in the fossil record every billion years
B. changes in the fossil record every million years
C. gradual changes in the fossil record
D. sudden changes in the fossil record

3 Which is NOT a cause of a mass extinction event?

A. meteorite collision
B. severe hurricane
C. tectonic activity
D. volcanic activity

4 Which is the correct order of eras, from oldest to youngest?

A. Cenozoic, Mesozoic, Paleozoic
B. Mesozoic, Cenozoic, Paleozoic
C. Paleozoic, Cenozoic, Mesozoic
D. Paleozoic, Mesozoic, Cenozoic

5 Which were the first organisms to inhabit land environments?

A. amphibians
B. plants
C. reptiles
D. trilobites

6 Which event(s) produced the Appalachian Mountains?

A. breakup of Pangaea
B. collisions of continents
C. flooding of the continent
D. opening of the Atlantic Ocean

7 Which was NOT associated with the Mesozoic era?

A. *Archaeopteryx*
B. plesiosaurs
C. pterosaurs
D. *Tiktaalik*

8 Which is true for the beginning of the Cenozoic era?

A. Mammals and dinosaurs lived together.
B. Mammals first evolved.
C. Dinosaurs had killed all mammals.
D. Dinosaurs were extinct.

9 What is unrealistic about the picture on this stamp?

A. Dinosaurs were not this large.
B. Dinosaurs did not have long necks.
C. Humans did not live with dinosaurs.
D. Early humans did not use stone tools.

Critical Thinking

10 **Hypothesize** how a major change in global climate could lead to a mass extinction.

11 **Evaluate** how the Permian-Triassic mass extinction affected the evolution of life.

12 **Predict** what Earth's climate might be like if sea level were very low.

13 **Differentiate** between amphibians and reptiles. What feature enabled reptiles—but not amphibians—to be successful on land?

14 **Hypothesize** how the bone structure of dinosaur limbs might have contributed to the success of dinosaurs during the Mesozoic era.

15 **Debate** Some scientists argue that humans have changed Earth so much that a new epoch—the Anthropocene epoch—should be added to the geologic time scale. Explain whether you think this is a good idea and, if so, when it should begin.

16 **Interpret Graphics** What is wrong with the geologic time line shown below?

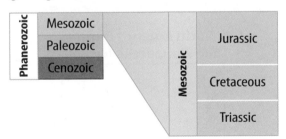

Writing in Science

17 **Decide** which period of Earth's history you would want to visit if you could travel back in time. Write a letter to a friend about your visit, describing the climate, the organisms, and the positions of Earth's continents at the time of your visit. Include a main idea, supporting details and examples, and a concluding sentence.

REVIEW THE BIG IDEA

18 What have scientists learned about Earth's past by studying rocks and fossils? How is the evolution of Earth's life-forms affected by geologic events? Provide examples.

19 The photo below shows an extinct dinosaur. What changes on Earth can cause organisms to become extinct?

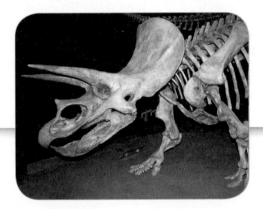

Math Skills

Review
— Math Practice —

Use Percentages

Use the table to answer the questions.

Era	Period	Epoch	Time Scale
Cenozoic	Quaternary	Holocene Pleistocene	10,000 years ago 1.8 mya
	Tertiary	Pliocene	5.3 mya
		Miocene	23.8 mya
		Oligocene	33.7 mya
		Eocene	54.8 mya
		Paleocene	65.5 mya

20 What percentage of the Quaternary period is represented by the Holocene epoch?

21 What percentage of the Tertiary period is represented by the Pliocene epoch?

Record your answers on the answer sheet provided by your teacher or on a sheet of paper.

Multiple Choice

Use the figure below to answer question 1.

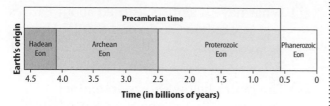

1 Approximately how long did Precambrian time last?

   A 0.5 billion years

   B 3.5 billion years

   C 4.0 billion years

   D 4.25 billion years

2 Which is the smallest unit of geologic time?

   A eon

   B epoch

   C era

   D period

3 Which is known as the age of invertebrates?

   A Early Cenozoic

   B Early Paleozoic

   C Late Mesozoic

   D Late Precambrian

4 Which made dinosaurs different from modern-day reptiles?

   A head shape

   B hip structure

   C jaw alignment

   D tail length

5 What is the approximate age of the oldest fossils of early human ancestors?

   A 10,000 years

   B 6 million years

   C 65 million years

   D 1.5 billion years

6 Which was NOT an adaptation that enabled amphibians to live on land?

   A ability to breathe oxygen

   B ability to lay eggs on land

   C strong limbs

   D thick skin

7 Which is considered a mega-mammal?

   A Archaeopteryx

   B plesiosaur

   C Tiktaalik

   D woolly mammoth

Use the figure below to answer question 8.

North America During the Pleistocene Ice Age

8 The figure above is a map of glacial coverage in North America. Which section of the United States would most likely have the greatest number of glacial grooves?

   A the Northeast

   B the Northwest

   C the Southeast

   D the Southwest

Use the graph below to answer question 9.

Sea Level Rise During Mesozoic

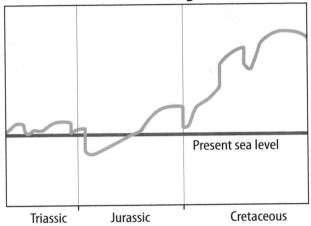

Present sea level

Triassic   Jurassic   Cretaceous

9 Based on the graph above, when might inland seas have covered much of Earth's continents?

   A Early Cretaceous

   B Early Jurassic

   C Middle Triassic

   D Late Cretaceous

10 Which did NOT occur in the Paleozoic era?

   A appearance of mammals

   B development of coal swamps

   C evolution of invertebrates

   D formation of Pangaea

11 What do geologists use to mark divisions in geologic time?

   A abrupt changes in the fossil record

   B frequent episodes of climate change

   C movements of Earth's tectonic plates

   D rates of radioactive mineral decay

Constructed Response

Use the graph below to answer questions 12 and 13.

Extinction Events

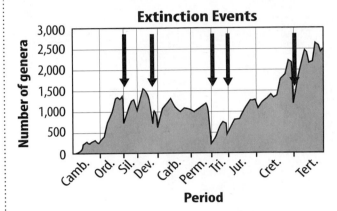

Period

12 In the graph above, what events do the arrows mark? What happens during these events?

13 What event appears to have had the greatest impact? Explain your answer in terms of the graph.

14 What are two possible reasons why large populations of organisms die?

15 What is the relationship between the evolution of marsupials and the movement of Earth's tectonic plates?

16 Why did new and existing aquatic organisms flourish during the Mesozoic era? Use the terms *glaciers*, *Pangaea*, and *sea level* in your explanation.

17 What is the link between iridium and the mass extinction of dinosaurs?

NEED EXTRA HELP?																	
If You Missed Question...	1	2	3	4	5	6	7	8	9	10	11	12	13	14	15	16	17
Go to Lesson...	1	1	2	3	4	2	4	4	3	2	1	1	1	1-3	4	3	1

Student Resources

For Students and Parents/Guardians

These resources are designed to help you achieve success in science. You will find useful information on laboratory safety, math skills, and science skills. In addition, science reference materials are found in the Reference Handbook. You'll find the information you need to learn and sharpen your skills in these resources.

Table of Contents

Scientific Methods

Scientists use an orderly approach called the scientific method to solve problems. This includes organizing and recording data so others can understand them. Scientists use many variations in this method when they solve problems.

Identify a Question

The first step in a scientific investigation or experiment is to identify a question to be answered or a problem to be solved. For example, you might ask which gasoline is the most efficient.

Gather and Organize Information

After you have identified your question, begin gathering and organizing information. There are many ways to gather information, such as researching in a library, interviewing those knowledgeable about the subject, and testing and working in the laboratory and field. Fieldwork is investigations and observations done outside of a laboratory.

Researching Information Before moving in a new direction, it is important to gather the information that already is known about the subject. Start by asking yourself questions to determine exactly what you need to know. Then you will look for the information in various reference sources, like the student is doing in **Figure 1.** Some sources may include textbooks, encyclopedias, government documents, professional journals, science magazines, and the Internet. Always list the sources of your information.

Figure 1 The Internet can be a valuable research tool.

Evaluate Sources of Information Not all sources of information are reliable. You should evaluate all of your sources of information, and use only those you know to be dependable. For example, if you are researching ways to make homes more energy efficient, a site written by the U.S. Department of Energy would be more reliable than a site written by a company that is trying to sell a new type of weatherproofing material. Also, remember that research always is changing. Consult the most current resources available to you. For example, a 1985 resource about saving energy would not reflect the most recent findings.

Sometimes scientists use data that they did not collect themselves, or conclusions drawn by other researchers. This data must be evaluated carefully. Ask questions about how the data were obtained, if the investigation was carried out properly, and if it has been duplicated exactly with the same results. Would you reach the same conclusion from the data? Only when you have confidence in the data can you believe it is true and feel comfortable using it.

SCIENCE SKILL HANDBOOK

MATH SKILL HANDBOOK

FOLDABLES HANDBOOK

REFERENCE HANDBOOK

GLOSSARY/ GLOSARIO

INDEX

Interpret Scientific Illustrations As you research a topic in science, you will see drawings, diagrams, and photographs to help you understand what you read. Some illustrations are included to help you understand an idea that you can't see easily by yourself, like the tiny particles in an atom in **Figure 2.** A drawing helps many people to remember details more easily and provides examples that clarify difficult concepts or give additional information about the topic you are studying. Most illustrations have labels or a caption to identify or to provide more information.

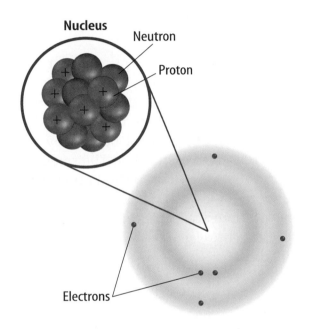

Figure 2 This drawing shows an atom of carbon with its six protons, six neutrons, and six electrons.

Concept Maps One way to organize data is to draw a diagram that shows relationships among ideas (or concepts). A concept map can help make the meanings of ideas and terms more clear, and help you understand and remember what you are studying. Concept maps are useful for breaking large concepts down into smaller parts, making learning easier.

Network Tree A type of concept map that not only shows a relationship, but how the concepts are related is a network tree, shown in **Figure 3.** In a network tree, the words are written in the ovals, while the description of the type of relationship is written across the connecting lines.

When constructing a network tree, write down the topic and all major topics on separate pieces of paper or notecards. Then arrange them in order from general to specific. Branch the related concepts from the major concept and describe the relationship on the connecting line. Continue to more specific concepts until finished.

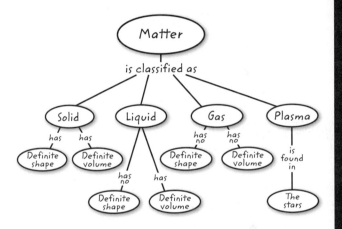

Figure 3 A network tree shows how concepts or objects are related.

Events Chain Another type of concept map is an events chain. Sometimes called a flow chart, it models the order or sequence of items. An events chain can be used to describe a sequence of events, the steps in a procedure, or the stages of a process.

When making an events chain, first find the one event that starts the chain. This event is called the initiating event. Then, find the next event and continue until the outcome is reached, as shown in **Figure 4** on the next page.

SCIENCE SKILL HANDBOOK

MATH SKILL HANDBOOK

FOLDABLES HANDBOOK

REFERENCE HANDBOOK

GLOSSARY/ GLOSARIO

INDEX

SCIENCE SKILL HANDBOOK

MATH SKILL HANDBOOK

FOLDABLES HANDBOOK

REFERENCE HANDBOOK

GLOSSARY/ GLOSARIO

INDEX

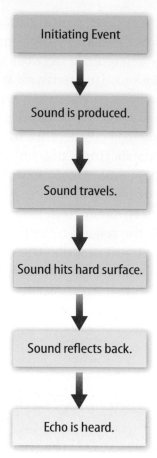

Figure 4 Events-chain concept maps show the order of steps in a process or event. This concept map shows how a sound makes an echo.

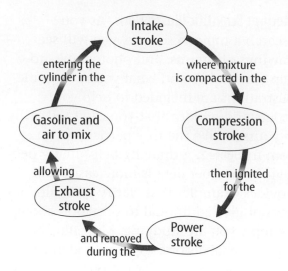

Figure 5 A cycle map shows events that occur in a cycle.

Spider Map A type of concept map that you can use for brainstorming is the spider map. When you have a central idea, you might find that you have a jumble of ideas that relate to it but are not necessarily clearly related to each other. The spider map on sound in **Figure 6** shows that if you write these ideas outside the main concept, then you can begin to separate and group unrelated terms so they become more useful.

Cycle Map A specific type of events chain is a cycle map. It is used when the series of events do not produce a final outcome, but instead relate back to the beginning event, such as in **Figure 5.** Therefore, the cycle repeats itself.

To make a cycle map, first decide what event is the beginning event. This is also called the initiating event. Then list the next events in the order that they occur, with the last event relating back to the initiating event. Words can be written between the events that describe what happens from one event to the next. The number of events in a cycle map can vary, but usually contain three or more events.

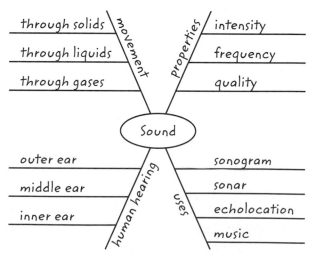

Figure 6 A spider map allows you to list ideas that relate to a central topic but not necessarily to one another.

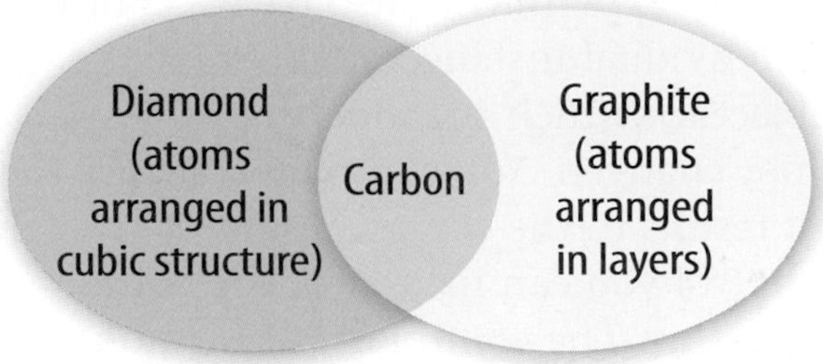

Figure 7 This Venn diagram compares and contrasts two substances made from carbon.

Venn Diagram To illustrate how two subjects compare and contrast you can use a Venn diagram. You can see the characteristics that the subjects have in common and those that they do not, shown in **Figure 7.**

To create a Venn diagram, draw two overlapping ovals that are big enough to write in. List the characteristics unique to one subject in one oval, and the characteristics of the other subject in the other oval. The characteristics in common are listed in the overlapping section.

Make and Use Tables One way to organize information so it is easier to understand is to use a table. Tables can contain numbers, words, or both.

To make a table, list the items to be compared in the first column and the characteristics to be compared in the first row. The title should clearly indicate the content of the table, and the column or row heads should be clear. Notice that in **Table 1** the units are included.

Table 1 Recyclables Collected During Week			
Day of Week	Paper (kg)	Aluminum (kg)	Glass (kg)
Monday	5.0	4.0	12.0
Wednesday	4.0	1.0	10.0
Friday	2.5	2.0	10.0

Make a Model One way to help you better understand the parts of a structure, the way a process works, or to show things too large or small for viewing is to make a model. For example, an atomic model made of a plastic-ball nucleus and chenille stem electron shells can help you visualize how the parts of an atom relate to each other. Other types of models can be devised on a computer or represented by equations.

Form a Hypothesis

A possible explanation based on previous knowledge and observations is called a hypothesis. After researching gasoline types and recalling previous experiences in your family's car you form a hypothesis—our car runs more efficiently because we use premium gasoline. To be valid, a hypothesis has to be something you can test by using an investigation.

Predict When you apply a hypothesis to a specific situation, you predict something about that situation. A prediction makes a statement in advance, based on prior observation, experience, or scientific reasoning. People use predictions to make everyday decisions. Scientists test predictions by performing investigations. Based on previous observations and experiences, you might form a prediction that cars are more efficient with premium gasoline. The prediction can be tested in an investigation.

Design an Experiment A scientist needs to make many decisions before beginning an investigation. Some of these include: how to carry out the investigation, what steps to follow, how to record the data, and how the investigation will answer the question. It also is important to address any safety concerns.

SCIENCE SKILL HANDBOOK

MATH SKILL HANDBOOK

FOLDABLES HANDBOOK

REFERENCE HANDBOOK

GLOSSARY/ GLOSARIO

INDEX

Test the Hypothesis

Now that you have formed your hypothesis, you need to test it. Using an investigation, you will make observations and collect data, or information. This data might either support or not support your hypothesis. Scientists collect and organize data as numbers and descriptions.

Follow a Procedure In order to know what materials to use, as well as how and in what order to use them, you must follow a procedure. **Figure 8** shows a procedure you might follow to test your hypothesis.

Procedure

Step 1 Use regular gasoline for two weeks.

Step 2 Record the number of kilometers between fill-ups and the amount of gasoline used.

Step 3 Switch to premium gasoline for two weeks.

Step 4 Record the number of kilometers between fill-ups and the amount of gasoline used.

Figure 8 A procedure tells you what to do step-by-step.

Identify and Manipulate Variables and Controls

In any experiment, it is important to keep everything the same except for the item you are testing. The one factor you change is called the independent variable. The change that results is the dependent variable. Make sure you have only one independent variable, to assure yourself of the cause of the changes you observe in the dependent variable. For example, in your gasoline experiment the type of fuel is the independent variable. The dependent variable is the efficiency.

Many experiments also have a control—an individual instance or experimental subject for which the independent variable is not changed. You can then compare the test results to the control results. To design a control you can have two cars of the same type. The control car uses regular gasoline for four weeks. After you are done with the test, you can compare the experimental results to the control results.

Collect Data

Whether you are carrying out an investigation or a short observational experiment, you will collect data, as shown in **Figure 9.** Scientists collect data as numbers and descriptions and organize them in specific ways.

Observe Scientists observe items and events, then record what they see. When they use only words to describe an observation, it is called qualitative data. Scientists' observations also can describe how much there is of something. These observations use numbers, as well as words, in the description and are called quantitative data. For example, if a sample of the element gold is described as being "shiny and very dense" the data are qualitative. Quantitative data on this sample of gold might include "a mass of 30 g and a density of 19.3 g/cm^3."

Figure 9 Collecting data is one way to gather information directly.

Figure 10 Record data neatly and clearly so it is easy to understand.

When you make observations you should examine the entire object or situation first, and then look carefully for details. It is important to record observations accurately and completely. Always record your notes immediately as you make them, so you do not miss details or make a mistake when recording results from memory. Never put unidentified observations on scraps of paper. Instead they should be recorded in a notebook, like the one in **Figure 10.** Write your data neatly so you can easily read it later. At each point in the experiment, record your observations and label them. That way, you will not have to determine what the figures mean when you look at your notes later. Set up any tables that you will need to use ahead of time, so you can record any observations right away. Remember to avoid bias when collecting data by not including personal thoughts when you record observations. Record only what you observe.

Estimate Scientific work also involves estimating. To estimate is to make a judgment about the size or the number of something without measuring or counting. This is important when the number or size of an object or population is too large or too difficult to accurately count or measure.

Sample Scientists may use a sample or a portion of the total number as a type of estimation. To sample is to take a small, representative portion of the objects or organisms of a population for research. By making careful observations or manipulating variables within that portion of the group, information is discovered and conclusions are drawn that might apply to the whole population. A poorly chosen sample can be unrepresentative of the whole. If you were trying to determine the rainfall in an area, it would not be best to take a rainfall sample from under a tree.

Measure You use measurements every day. Scientists also take measurements when collecting data. When taking measurements, it is important to know how to use measuring tools properly. Accuracy also is important.

Length To measure length, the distance between two points, scientists use meters. Smaller measurements might be measured in centimeters or millimeters.

Length is measured using a metric ruler or meterstick. When using a metric ruler, line up the 0-cm mark with the end of the object being measured and read the number of the unit where the object ends. Look at the metric ruler shown in **Figure 11.** The centimeter lines are the long, numbered lines, and the shorter lines are millimeter lines. In this instance, the length would be 4.50 cm.

Figure 11 This metric ruler has centimeter and millimeter divisions.

SCIENCE SKILL HANDBOOK

MATH SKILL HANDBOOK

FOLDABLES HANDBOOK

REFERENCE HANDBOOK

GLOSSARY/ GLOSARIO

INDEX

SCIENCE SKILL HANDBOOK

MATH SKILL HANDBOOK

FOLDABLES HANDBOOK

REFERENCE HANDBOOK

GLOSSARY/ GLOSARIO

INDEX

Mass The SI unit for mass is the kilogram (kg). Scientists can measure mass using units formed by adding metric prefixes to the unit gram (g), such as milligram (mg). To measure mass, you might use a triple-beam balance similar to the one shown in **Figure 12.** The balance has a pan on one side and a set of beams on the other side. Each beam has a rider that slides on the beam.

When using a triple-beam balance, place an object on the pan. Slide the largest rider along its beam until the pointer drops below zero. Then move it back one notch. Repeat the process for each rider proceeding from the larger to smaller until the pointer swings an equal distance above and below the zero point. Sum the masses on each beam to find the mass of the object. Move all riders back to zero when finished.

Instead of putting materials directly on the balance, scientists often take a tare of a container. A tare is the mass of a container into which objects or substances are placed for measuring their masses. To find the mass of objects or substances, find the mass of a clean container. Remove the container from the pan, and place the object or substances in the container. Find the mass of the container with the materials in it. Subtract the mass of the empty container from the mass of the filled container to find the mass of the materials you are using.

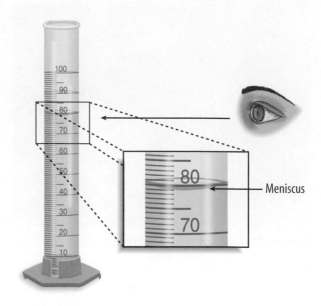

Figure 13 Graduated cylinders measure liquid volume.

Liquid Volume To measure liquids, the unit used is the liter. When a smaller unit is needed, scientists might use a milliliter. Because a milliliter takes up the volume of a cube measuring 1 cm on each side it also can be called a cubic centimeter ($cm^3 = cm \times cm \times cm$).

You can use beakers and graduated cylinders to measure liquid volume. A graduated cylinder, shown in **Figure 13,** is marked from bottom to top in milliliters. In lab, you might use a 10-mL graduated cylinder or a 100-mL graduated cylinder. When measuring liquids, notice that the liquid has a curved surface. Look at the surface at eye level, and measure the bottom of the curve. This is called the meniscus. The graduated cylinder in **Figure 13** contains 79.0 mL, or 79.0 cm^3, of a liquid.

Temperature Scientists often measure temperature using the Celsius scale. Pure water has a freezing point of 0°C and boiling point of 100°C. The unit of measurement is degrees Celsius. Two other scales often used are the Fahrenheit and Kelvin scales.

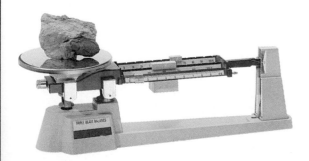

Figure 12 A triple-beam balance is used to determine the mass of an object.

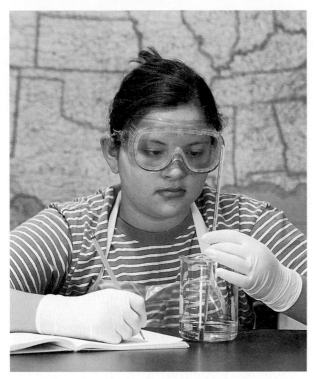

Figure 14 A thermometer measures the temperature of an object.

Scientists use a thermometer to measure temperature. Most thermometers in a laboratory are glass tubes with a bulb at the bottom end containing a liquid such as colored alcohol. The liquid rises or falls with a change in temperature. To read a glass thermometer like the thermometer in **Figure 14,** rotate it slowly until a red line appears. Read the temperature where the red line ends.

Form Operational Definitions
An operational definition defines an object by how it functions, works, or behaves. For example, when you are playing hide and seek and a tree is home base, you have created an operational definition for a tree.

Objects can have more than one operational definition. For example, a ruler can be defined as a tool that measures the length of an object (how it is used). It can also be a tool with a series of marks used as a standard when measuring (how it works).

Analyze the Data

To determine the meaning of your observations and investigation results, you will need to look for patterns in the data. Then you must think critically to determine what the data mean. Scientists use several approaches when they analyze the data they have collected and recorded. Each approach is useful for identifying specific patterns.

Interpret Data
The word *interpret* means "to explain the meaning of something." When analyzing data from an experiment, try to find out what the data show. Identify the control group and the test group to see whether changes in the independent variable have had an effect. Look for differences in the dependent variable between the control and test groups.

Classify
Sorting objects or events into groups based on common features is called classifying. When classifying, first observe the objects or events to be classified. Then select one feature that is shared by some members in the group, but not by all. Place those members that share that feature in a subgroup. You can classify members into smaller and smaller subgroups based on characteristics. Remember that when you classify, you are grouping objects or events for a purpose. Keep your purpose in mind as you select the features to form groups and subgroups.

Compare and Contrast
Observations can be analyzed by noting the similarities and differences between two or more objects or events that you observe. When you look at objects or events to see how they are similar, you are comparing them. Contrasting is looking for differences in objects or events.

SCIENCE SKILL HANDBOOK
MATH SKILL HANDBOOK
FOLDABLES HANDBOOK
REFERENCE HANDBOOK
GLOSSARY/ GLOSARIO
INDEX

SCIENCE SKILL HANDBOOK

MATH SKILL HANDBOOK

FOLDABLES HANDBOOK

REFERENCE HANDBOOK

GLOSSARY/ GLOSARIO

INDEX

Recognize Cause and Effect A cause is a reason for an action or condition. The effect is that action or condition. When two events happen together, it is not necessarily true that one event caused the other. Scientists must design a controlled investigation to recognize the exact cause and effect.

Draw Conclusions

When scientists have analyzed the data they collected, they proceed to draw conclusions about the data. These conclusions are sometimes stated in words similar to the hypothesis that you formed earlier. They may confirm a hypothesis, or lead you to a new hypothesis.

Infer Scientists often make inferences based on their observations. An inference is an attempt to explain observations or to indicate a cause. An inference is not a fact, but a logical conclusion that needs further investigation. For example, you may infer that a fire has caused smoke. Until you investigate, however, you do not know for sure.

Apply When you draw a conclusion, you must apply those conclusions to determine whether the data supports the hypothesis. If your data do not support your hypothesis, it does not mean that the hypothesis is wrong. It means only that the result of the investigation did not support the hypothesis. Maybe the experiment needs to be redesigned, or some of the initial observations on which the hypothesis was based were incomplete or biased. Perhaps more observation or research is needed to refine your hypothesis. A successful investigation does not always come out the way you originally predicted.

Avoid Bias Sometimes a scientific investigation involves making judgments. When you make a judgment, you form an opinion. It is important to be honest and not to allow any expectations of results to bias your judgments. This is important throughout the entire investigation, from researching to collecting data to drawing conclusions.

Communicate

The communication of ideas is an important part of the work of scientists. A discovery that is not reported will not advance the scientific community's understanding or knowledge. Communication among scientists also is important as a way of improving their investigations.

Scientists communicate in many ways, from writing articles in journals and magazines that explain their investigations and experiments, to announcing important discoveries on television and radio. Scientists also share ideas with colleagues on the Internet or present them as lectures, like the student is doing in **Figure 15.**

Figure 15 A student communicates to his peers about his investigation.

These safety symbols are used in laboratory and field investigations in this book to indicate possible hazards. Learn the meaning of each symbol and refer to this page often. *Remember to wash your hands thoroughly after completing lab procedures.*

PROTECTIVE EQUIPMENT   Do not begin any lab without the proper protection equipment.

 GOGGLES Proper eye protection must be worn when performing or observing science activities that involve items or conditions as listed below.

 APRON Wear an approved apron when using substances that could stain, wet, or destroy cloth.

 SOAP Wash hands with soap and water before removing goggles and after all lab activities.

 GLOVES Wear gloves when working with biological materials, chemicals, animals, or materials that can stain or irritate hands.

LABORATORY HAZARDS

Symbols	Potential Hazards	Precaution	Response
DISPOSAL	contamination of classroom or environment due to improper disposal of materials such as chemicals and live specimens	• DO NOT dispose of hazardous materials in the sink or trash can. • Dispose of wastes as directed by your teacher.	• If hazardous materials are disposed of improperly, notify your teacher immediately.
EXTREME TEMPERATURE	skin burns due to extremely hot or cold materials such as hot glass, liquids, or metals; liquid nitrogen; dry ice	• Use proper protective equipment, such as hot mitts and/or tongs, when handling objects with extreme temperatures.	• If injury occurs, notify your teacher immediately.
SHARP OBJECTS	punctures or cuts from sharp objects such as razor blades, pins, scalpels, and broken glass	• Handle glassware carefully to avoid breakage. • Walk with sharp objects pointed downward, away from you and others.	• If broken glass or injury occurs, notify your teacher immediately.
ELECTRICAL	electric shock or skin burn due to improper grounding, short circuits, liquid spills, or exposed wires	• Check condition of wires and apparatus for fraying or uninsulated wires, and broken or cracked equipment. • Use only GFCI-protected outlets	• DO NOT attempt to fix electrical problems. Notify your teacher immediately.
CHEMICAL	skin irritation or burns, breathing difficulty, and/or poisoning due to touching, swallowing, or inhalation of chemicals such as acids, bases, bleach, metal compounds, iodine, poinsettias, pollen, ammonia, acetone, nail polish remover, heated chemicals, mothballs, and any other chemicals labeled or known to be dangerous	• Wear proper protective equipment such as goggles, apron, and gloves when using chemicals. • Ensure proper room ventilation or use a fume hood when using materials that produce fumes. • NEVER smell fumes directly. • NEVER taste or eat any material in the laboratory.	• If contact occurs, immediately flush affected area with water and notify your teacher. • If a spill occurs, leave the area immediately and notify your teacher.
FLAMMABLE	unexpected fire due to liquids or gases that ignite easily such as rubbing alcohol	• Avoid open flames, sparks, or heat when flammable liquids are present.	• If a fire occurs, leave the area immediately and notify your teacher.
OPEN FLAME	burns or fire due to open flame from matches, Bunsen burners, or burning materials	• Tie back loose hair and clothing. • Keep flame away from all materials. • Follow teacher instructions when lighting and extinguishing flames. • Use proper protection, such as hot mitts or tongs, when handling hot objects.	• If a fire occurs, leave the area immediately and notify your teacher.
ANIMAL SAFETY	injury to or from laboratory animals	• Wear proper protective equipment such as gloves, apron, and goggles when working with animals. • Wash hands after handling animals.	• If injury occurs, notify your teacher immediately.
BIOLOGICAL	infection or adverse reaction due to contact with organisms such as bacteria, fungi, and biological materials such as blood, animal or plant materials	• Wear proper protective equipment such as gloves, goggles, and apron when working with biological materials. • Avoid skin contact with an organism or any part of the organism. • Wash hands after handling organisms.	• If contact occurs, wash the affected area and notify your teacher immediately.
FUME	breathing difficulties from inhalation of fumes from substances such as ammonia, acetone, nail polish remover, heated chemicals, and mothballs	• Wear goggles, apron, and gloves. • Ensure proper room ventilation or use a fume hood when using substances that produce fumes. • NEVER smell fumes directly.	• If a spill occurs, leave area and notify your teacher immediately.
IRRITANT	irritation of skin, mucous membranes, or respiratory tract due to materials such as acids, bases, bleach, pollen, mothballs, steel wool, and potassium permanganate	• Wear goggles, apron, and gloves. • Wear a dust mask to protect against fine particles.	• If skin contact occurs, immediately flush the affected area with water and notify your teacher.
RADIOACTIVE	excessive exposure from alpha, beta, and gamma particles	• Remove gloves and wash hands with soap and water before removing remainder of protective equipment.	• If cracks or holes are found in the container, notify your teacher immediately.

SCIENCE SKILL HANDBOOK

MATH SKILL HANDBOOK

FOLDABLES HANDBOOK

REFERENCE HANDBOOK

GLOSSARY/ GLOSARIO

INDEX

Safety in the Science Laboratory

Introduction to Science Safety

The science laboratory is a safe place to work if you follow standard safety procedures. Being responsible for your own safety helps to make the entire laboratory a safer place for everyone. When performing any lab, read and apply the caution statements and safety symbol listed at the beginning of the lab.

General Safety Rules

1. Complete the *Lab Safety Form* or other safety contract BEFORE starting any science lab.

2. Study the procedure. Ask your teacher any questions. Be sure you understand safety symbols shown on the page.

3. Notify your teacher about allergies or other health conditions that can affect your participation in a lab.

4. Learn and follow use and safety procedures for your equipment. If unsure, ask your teacher.

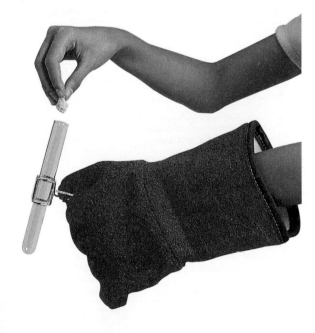

5. Never eat, drink, chew gum, apply cosmetics, or do any personal grooming in the lab. Never use lab glassware as food or drink containers. Keep your hands away from your face and mouth.

6. Know the location and proper use of the safety shower, eye wash, fire blanket, and fire alarm.

Prevent Accidents

1. Use the safety equipment provided to you. Goggles and a safety apron should be worn during investigations.

2. Do NOT use hair spray, mousse, or other flammable hair products. Tie back long hair and tie down loose clothing.

3. Do NOT wear sandals or other open-toed shoes in the lab.

4. Remove jewelry on hands and wrists. Loose jewelry, such as chains and long necklaces, should be removed to prevent them from getting caught in equipment.

5. Do not taste any substances or draw any material into a tube with your mouth.

6. Proper behavior is expected in the lab. Practical jokes and fooling around can lead to accidents and injury.

7. Keep your work area uncluttered.

Laboratory Work

1. Collect and carry all equipment and materials to your work area before beginning a lab.

2. Remain in your own work area unless given permission by your teacher to leave it.

SCIENCE SKILL HANDBOOK

MATH SKILL HANDBOOK

FOLDABLES HANDBOOK

REFERENCE HANDBOOK

GLOSSARY/ GLOSARIO

INDEX

3. Always slant test tubes away from yourself and others when heating them, adding substances to them, or rinsing them.

4. If instructed to smell a substance in a container, hold the container a short distance away and fan vapors toward your nose.

5. Do NOT substitute other chemicals/substances for those in the materials list unless instructed to do so by your teacher.

6. Do NOT take any materials or chemicals outside of the laboratory.

7. Stay out of storage areas unless instructed to be there and supervised by your teacher.

Laboratory Cleanup

1. Turn off all burners, water, and gas, and disconnect all electrical devices.

2. Clean all pieces of equipment and return all materials to their proper places.

3. Dispose of chemicals and other materials as directed by your teacher. Place broken glass and solid substances in the proper containers. Never discard materials in the sink.

4. Clean your work area.

5. Wash your hands with soap and water thoroughly BEFORE removing your goggles.

Emergencies

1. Report any fire, electrical shock, glassware breakage, spill, or injury, no matter how small, to your teacher immediately. Follow his or her instructions.

2. If your clothing should catch fire, STOP, DROP, and ROLL. If possible, smother it with the fire blanket or get under a safety shower. NEVER RUN.

3. If a fire should occur, turn off all gas and leave the room according to established procedures.

4. In most instances, your teacher will clean up spills. Do NOT attempt to clean up spills unless you are given permission and instructions to do so.

5. If chemicals come into contact with your eyes or skin, notify your teacher immediately. Use the eyewash, or flush your skin or eyes with large quantities of water.

6. The fire extinguisher and first-aid kit should only be used by your teacher unless it is an extreme emergency and you have been given permission.

7. If someone is injured or becomes ill, only a professional medical provider or someone certified in first aid should perform first-aid procedures.

SCIENCE SKILL HANDBOOK

MATH SKILL HANDBOOK

FOLDABLES HANDBOOK

REFERENCE HANDBOOK

GLOSSARY/ GLOSARIO

INDEX

Use Fractions

A fraction compares a part to a whole. In the fraction $\frac{2}{3}$, the 2 represents the part and is the numerator. The 3 represents the whole and is the denominator.

Reduce Fractions To reduce a fraction, you must find the largest factor that is common to both the numerator and the denominator, the greatest common factor (GCF). Divide both numbers by the GCF. The fraction has then been reduced, or it is in its simplest form.

Example

Twelve of the 20 chemicals in the science lab are in powder form. What fraction of the chemicals used in the lab are in powder form?

Step 1 Write the fraction.

$$\frac{part}{whole} = \frac{12}{20}$$

Step 2 To find the GCF of the numerator and denominator, list all of the factors of each number.

Factors of 12: 1, 2, 3, 4, 6, 12 (the numbers that divide evenly into 12)

Factors of 20: 1, 2, 4, 5, 10, 20 (the numbers that divide evenly into 20)

Step 3 List the common factors.

1, 2, 4

Step 4 Choose the greatest factor in the list. The GCF of 12 and 20 is 4.

Step 5 Divide the numerator and denominator by the GCF.

$$\frac{12 \div 4}{20 \div 4} = \frac{3}{5}$$

In the lab, $\frac{3}{5}$ of the chemicals are in powder form.

Practice Problem At an amusement park, 66 of 90 rides have a height restriction. What fraction of the rides, in its simplest form, has a height restriction?

Add and Subtract Fractions with Like Denominators To add or subtract fractions with the same denominator, add or subtract the numerators and write the sum or difference over the denominator. After finding the sum or difference, find the simplest form for your fraction.

Example 1

In the forest outside your house, $\frac{1}{8}$ of the animals are rabbits, $\frac{3}{8}$ are squirrels, and the remainder are birds and insects. How many are mammals?

Step 1 Add the numerators.

$$\frac{1}{8} + \frac{3}{8} = \frac{(1 + 3)}{8} = \frac{4}{8}$$

Step 2 Find the GCF.

$\frac{4}{8}$ (GCF, 4)

Step 3 Divide the numerator and denominator by the GCF.

$$\frac{4 \div 4}{8 \div 4} = \frac{1}{2}$$

$\frac{1}{2}$ of the animals are mammals.

Example 2

If $\frac{7}{16}$ of the Earth is covered by freshwater, and $\frac{1}{16}$ of that is in glaciers, how much freshwater is not frozen?

Step 1 Subtract the numerators.

$$\frac{7}{16} - \frac{1}{16} = \frac{(7 - 1)}{16} = \frac{6}{16}$$

Step 2 Find the GCF.

$\frac{6}{16}$ (GCF, 2)

Step 3 Divide the numerator and denominator by the GCF.

$$\frac{6 \div 2}{16 \div 2} = \frac{3}{8}$$

$\frac{3}{8}$ of the freshwater is not frozen.

Practice Problem A bicycle rider is riding at a rate of 15 km/h for $\frac{4}{9}$ of his ride, 10 km/h for $\frac{2}{9}$ of his ride, and 8 km/h for the remainder of the ride. How much of his ride is he riding at a rate greater than 8 km/h?

Add and Subtract Fractions with Unlike Denominators To add or subtract fractions with unlike denominators, first find the least common denominator (LCD). This is the smallest number that is a common multiple of both denominators. Rename each fraction with the LCD, and then add or subtract. Find the simplest form if necessary.

Example 1

A chemist makes a paste that is $\frac{1}{2}$ table salt (NaCl), $\frac{1}{3}$ sugar ($C_6H_{12}O_6$), and the remainder is water (H_2O). How much of the paste is a solid?

Step 1   Find the LCD of the fractions.

$$\frac{1}{2} + \frac{1}{3} \text{ (LCD, 6)}$$

Step 2   Rename each numerator and each denominator with the LCD.

Step 3   Add the numerators.

$$\frac{3}{6} + \frac{2}{6} = \frac{(3+2)}{6} = \frac{5}{6}$$

$\frac{5}{6}$ of the paste is a solid.

Example 2

The average precipitation in Grand Junction, CO, is $\frac{7}{10}$ inch in November, and $\frac{3}{5}$ inch in December. What is the total average precipitation?

Step 1   Find the LCD of the fractions.

$$\frac{7}{10} + \frac{3}{5} \text{ (LCD, 10)}$$

Step 2   Rename each numerator and each denominator with the LCD.

Step 3   Add the numerators.

$$\frac{7}{10} + \frac{6}{10} = \frac{(7+6)}{10} = \frac{13}{10}$$

$\frac{13}{10}$ inches total precipitation, or $1\frac{3}{10}$ inches.

Practice Problem On an electric bill, about $\frac{1}{8}$ of the energy is from solar energy and about $\frac{1}{10}$ is from wind power. How much of the total bill is from solar energy and wind power combined?

Example 3

In your body, $\frac{7}{10}$ of your muscle contractions are involuntary (cardiac and smooth muscle tissue). Smooth muscle makes $\frac{3}{15}$ of your muscle contractions. How many of your muscle contractions are made by cardiac muscle?

Step 1   Find the LCD of the fractions.

$$\frac{7}{10} - \frac{3}{15} \text{ (LCD, 30)}$$

Step 2   Rename each numerator and each denominator with the LCD.

$$\frac{7 \times 3}{10 \times 3} = \frac{21}{30}$$

$$\frac{3 \times 2}{15 \times 2} = \frac{6}{30}$$

Step 3   Subtract the numerators.

$$\frac{21}{30} - \frac{6}{30} = \frac{(21-6)}{30} = \frac{15}{30}$$

Step 4   Find the GCF.

$$\frac{15}{30} \text{ (GCF, 15)}$$

$$\frac{1}{2}$$

$\frac{1}{2}$ of all muscle contractions are cardiac muscle.

Example 4

Tony wants to make cookies that call for $\frac{3}{4}$ of a cup of flour, but he only has $\frac{1}{3}$ of a cup. How much more flour does he need?

Step 1   Find the LCD of the fractions.

$$\frac{3}{4} - \frac{1}{3} \text{ (LCD, 12)}$$

Step 2   Rename each numerator and each denominator with the LCD.

$$\frac{3 \times 3}{4 \times 3} = \frac{9}{12}$$

$$\frac{1 \times 4}{3 \times 4} = \frac{4}{12}$$

Step 3   Subtract the numerators.

$$\frac{9}{12} - \frac{4}{12} = \frac{(9-4)}{12} = \frac{5}{12}$$

$\frac{5}{12}$ of a cup of flour

Practice Problem Using the information provided to you in Example 3 above, determine how many muscle contractions are voluntary (skeletal muscle).

Multiply Fractions To multiply with fractions, multiply the numerators and multiply the denominators. Find the simplest form if necessary.

Example

Multiply $\frac{3}{5}$ by $\frac{1}{3}$.

Step 1 Multiply the numerators and denominators.

$$\frac{3}{5} \times \frac{1}{3} = \frac{(3 \times 1)}{(5 \times 3)} \; \frac{3}{15}$$

Step 2 Find the GCF.

$$\frac{3}{15} \; (\text{GCF, 3})$$

Step 3 Divide the numerator and denominator by the GCF.

$$\frac{3 \div 3}{15 \div 3} = \frac{1}{5}$$

$\frac{3}{5}$ multiplied by $\frac{1}{3}$ is $\frac{1}{5}$.

Practice Problem Multiply $\frac{3}{14}$ by $\frac{5}{16}$.

Find a Reciprocal Two numbers whose product is 1 are called multiplicative inverses, or reciprocals.

Example

Find the reciprocal of $\frac{3}{8}$.

Step 1 Inverse the fraction by putting the denominator on top and the numerator on the bottom.

$$\frac{8}{3}$$

The reciprocal of $\frac{3}{8}$ is $\frac{8}{3}$.

Practice Problem Find the reciprocal of $\frac{4}{9}$.

Divide Fractions To divide one fraction by another fraction, multiply the dividend by the reciprocal of the divisor. Find the simplest form if necessary.

Example 1

Divide $\frac{1}{9}$ by $\frac{1}{3}$.

Step 1 Find the reciprocal of the divisor.

The reciprocal of $\frac{1}{3}$ is $\frac{3}{1}$.

Step 2 Multiply the dividend by the reciprocal of the divisor.

$$\frac{\frac{1}{9}}{\frac{1}{3}} = \frac{1}{9} \times \frac{3}{1} = \frac{(1 \times 3)}{(9 \times 1)} = \frac{3}{9}$$

Step 3 Find the GCF.

$$\frac{3}{9} \; (\text{GCF, 3})$$

Step 4 Divide the numerator and denominator by the GCF.

$$\frac{3 \div 3}{9 \div 3} = \frac{1}{3}$$

$\frac{1}{9}$ divided by $\frac{1}{3}$ is $\frac{1}{3}$.

Example 2

Divide $\frac{3}{5}$ by $\frac{1}{4}$.

Step 1 Find the reciprocal of the divisor.

The reciprocal of $\frac{1}{4}$ is $\frac{4}{1}$.

Step 2 Multiply the dividend by the reciprocal of the divisor.

$$\frac{\frac{3}{5}}{\frac{1}{4}} = \frac{3}{5} \times \frac{4}{1} = \frac{(3 \times 4)}{(5 \times 1)} = \frac{12}{5}$$

$\frac{3}{5}$ divided by $\frac{1}{4}$ is $\frac{12}{5}$ or $2\frac{2}{5}$.

Practice Problem Divide $\frac{3}{11}$ by $\frac{7}{10}$.

Use Ratios

When you compare two numbers by division, you are using a ratio. Ratios can be written 3 to 5, 3:5, or $\frac{3}{5}$. Ratios, like fractions, also can be written in simplest form.

Ratios can represent one type of probability, called odds. This is a ratio that compares the number of ways a certain outcome occurs to the number of possible outcomes. For example, if you flip a coin 100 times, what are the odds that it will come up heads? There are two possible outcomes, heads or tails, so the odds of coming up heads are 50:100. Another way to say this is that 50 out of 100 times the coin will come up heads. In its simplest form, the ratio is 1:2.

Example 1

A chemical solution contains 40 g of salt and 64 g of baking soda. What is the ratio of salt to baking soda as a fraction in simplest form?

Step 1 Write the ratio as a fraction.
$$\frac{\text{salt}}{\text{baking soda}} = \frac{40}{64}$$

Step 2 Express the fraction in simplest form. The GCF of 40 and 64 is 8.
$$\frac{40}{64} = \frac{40 \div 8}{64 \div 8} = \frac{5}{8}$$

The ratio of salt to baking soda in the sample is 5:8.

Example 2

Sean rolls a 6-sided die 6 times. What are the odds that the side with a 3 will show?

Step 1 Write the ratio as a fraction.
$$\frac{\text{number of sides with a 3}}{\text{number of possible sides}} = \frac{1}{6}$$

Step 2 Multiply by the number of attempts.
$$\frac{1}{6} \times 6 \text{ attempts} = \frac{6}{6} \text{ attempts} = 1 \text{ attempt}$$

1 attempt out of 6 will show a 3.

Practice Problem Two metal rods measure 100 cm and 144 cm in length. What is the ratio of their lengths in simplest form?

Use Decimals

A fraction with a denominator that is a power of ten can be written as a decimal. For example, 0.27 means $\frac{27}{100}$. The decimal point separates the ones place from the tenths place.

Any fraction can be written as a decimal using division. For example, the fraction $\frac{5}{8}$ can be written as a decimal by dividing 5 by 8. Written as a decimal, it is 0.625.

Add or Subtract Decimals When adding and subtracting decimals, line up the decimal points before carrying out the operation.

Example 1

Find the sum of 47.68 and 7.80.

Step 1 Line up the decimal places when you write the numbers.

$$
\begin{array}{r}
47.68 \\
+\ 7.80 \\
\end{array}
$$

Step 2 Add the decimals.

$$
\begin{array}{r}
\overset{1\ 1}{47.68} \\
+\ 7.80 \\
\hline
55.48 \\
\end{array}
$$

The sum of 47.68 and 7.80 is 55.48.

Example 2

Find the difference of 42.17 and 15.85.

Step 1 Line up the decimal places when you write the number.

$$
\begin{array}{r}
42.17 \\
-15.85 \\
\end{array}
$$

Step 2 Subtract the decimals.

$$
\begin{array}{r}
\overset{3\ 11}{4\cancel{2}.17} \\
-15.85 \\
\hline
26.32 \\
\end{array}
$$

The difference of 42.17 and 15.85 is 26.32.

Practice Problem Find the sum of 1.245 and 3.842.

SCIENCE SKILL HANDBOOK

MATH SKILL HANDBOOK

FOLDABLES HANDBOOK

REFERENCE HANDBOOK

GLOSSARY/ GLOSARIO

INDEX

Multiply Decimals To multiply decimals, multiply the numbers like numbers without decimal points. Count the decimal places in each factor. The product will have the same number of decimal places as the sum of the decimal places in the factors.

Example

Multiply 2.4 by 5.9.

Step 1 Multiply the factors like two whole numbers.

$24 \times 59 = 1416$

Step 2 Find the sum of the number of decimal places in the factors. Each factor has one decimal place, for a sum of two decimal places.

Step 3 The product will have two decimal places.

14.16

The product of 2.4 and 5.9 is 14.16.

Practice Problem Multiply 4.6 by 2.2.

Divide Decimals When dividing decimals, change the divisor to a whole number. To do this, multiply both the divisor and the dividend by the same power of ten. Then place the decimal point in the quotient directly above the decimal point in the dividend. Then divide as you do with whole numbers.

Example

Divide 8.84 by 3.4.

Step 1 Multiply both factors by 10.

$3.4 \times 10 = 34, 8.84 \times 10 = 88.4$

Step 2 Divide 88.4 by 34.

$$
\begin{array}{r}
2.6 \\
34\overline{)88.4} \\
-68 \\
\hline
204 \\
-204 \\
\hline
0
\end{array}
$$

8.84 divided by 3.4 is 2.6.

Practice Problem Divide 75.6 by 3.6.

Use Proportions

An equation that shows that two ratios are equivalent is a proportion. The ratios $\frac{2}{4}$ and $\frac{5}{10}$ are equivalent, so they can be written as $\frac{2}{4} = \frac{5}{10}$. This equation is a proportion.

When two ratios form a proportion, the cross products are equal. To find the cross products in the proportion $\frac{2}{4} = \frac{5}{10}$, multiply the 2 and the 10, and the 4 and the 5. Therefore $2 \times 10 = 4 \times 5$, or $20 = 20$.

Because you know that both ratios are equal, you can use cross products to find a missing term in a proportion. This is known as solving the proportion.

Example

The heights of a tree and a pole are proportional to the lengths of their shadows. The tree casts a shadow of 24 m when a 6-m pole casts a shadow of 4 m. What is the height of the tree?

Step 1 Write a proportion.

$$\frac{\text{height of tree}}{\text{height of pole}} = \frac{\text{length of tree's shadow}}{\text{length of pole's shadow}}$$

Step 2 Substitute the known values into the proportion. Let h represent the unknown value, the height of the tree.

$$\frac{h}{6} \times \frac{24}{4}$$

Step 3 Find the cross products.

$$h \times 4 = 6 \times 24$$

Step 4 Simplify the equation.

$$4h \times 144$$

Step 5 Divide each side by 4.

$$\frac{4h}{4} \times \frac{144}{4}$$

$$h = 36$$

The height of the tree is 36 m.

Practice Problem The ratios of the weights of two objects on the Moon and on Earth are in proportion. A rock weighing 3 N on the Moon weighs 18 N on Earth. How much would a rock that weighs 5 N on the Moon weigh on Earth?

Use Percentages

The word *percent* means "out of one hundred." It is a ratio that compares a number to 100. Suppose you read that 77 percent of Earth's surface is covered by water. That is the same as reading that the fraction of Earth's surface covered by water is $\frac{77}{100}$. To express a fraction as a percent, first find the equivalent decimal for the fraction. Then, multiply the decimal by 100 and add the percent symbol.

Example 1

Express $\frac{13}{20}$ as a percent.

Step 1 Find the equivalent decimal for the fraction.

$$
\begin{array}{r}
0.65 \\
20\overline{)13.00} \\
\underline{12\;0} \\
1\;00 \\
\underline{1\;00} \\
0
\end{array}
$$

Step 2 Rewrite the fraction $\frac{13}{20}$ as 0.65.

Step 3 Multiply 0.65 by 100 and add the % symbol.

$$0.65 \times 100 = 65 = 65\%$$

So, $\frac{13}{20} = 65\%$.

This also can be solved as a proportion.

Example 2

Express $\frac{13}{20}$ as a percent.

Step 1 Write a proportion.

$$\frac{13}{20} = \frac{x}{100}$$

Step 2 Find the cross products.

$$1300 = 20x$$

Step 3 Divide each side by 20.

$$\frac{1300}{20} = \frac{20x}{20}$$

$$65\% = x$$

Practice Problem In one year, 73 of 365 days were rainy in one city. What percent of the days in that city were rainy?

Solve One-Step Equations

A statement that two expressions are equal is an equation. For example, $A = B$ is an equation that states that A is equal to B.

An equation is solved when a variable is replaced with a value that makes both sides of the equation equal. To make both sides equal the inverse operation is used. Addition and subtraction are inverses, and multiplication and division are inverses.

Example 1

Solve the equation $x - 10 = 35$.

Step 1 Find the solution by adding 10 to each side of the equation.

$$x - 10 = 35$$
$$x - 10 + 10 = 35 - 10$$
$$x = 45$$

Step 2 Check the solution.

$$x - 10 = 35$$
$$45 - 10 = 35$$
$$35 = 35$$

Both sides of the equation are equal, so $x = 45$.

Example 2

In the formula $a = bc$, find the value of c if $a = 20$ and $b = 2$.

Step 1 Rearrange the formula so the unknown value is by itself on one side of the equation by dividing both sides by b.

$$a = bc$$
$$\frac{a}{b} = \frac{bc}{b}$$
$$\frac{a}{b} = c$$

Step 2 Replace the variables a and b with the values that are given.

$$\frac{a}{b} = c$$
$$\frac{20}{2} = c$$
$$10 = c$$

Step 3 Check the solution.

$$a = bc$$
$$20 = 2 \times 10$$
$$20 = 20$$

Both sides of the equation are equal, so $c = 10$ is the solution when $a = 20$ and $b = 2$.

Practice Problem In the formula $h = gd$, find the value of d if $g = 12.3$ and $h = 17.4$.

SCIENCE SKILL HANDBOOK

MATH SKILL HANDBOOK

FOLDABLES HANDBOOK

REFERENCE HANDBOOK

GLOSSARY/ GLOSARIO

INDEX

Use Statistics

The branch of mathematics that deals with collecting, analyzing, and presenting data is statistics. In statistics, there are three common ways to summarize data with a single number—the mean, the median, and the mode.

The **mean** of a set of data is the arithmetic average. It is found by adding the numbers in the data set and dividing by the number of items in the set.

The **median** is the middle number in a set of data when the data are arranged in numerical order. If there were an even number of data points, the median would be the mean of the two middle numbers.

The **mode** of a set of data is the number or item that appears most often.

Another number that often is used to describe a set of data is the range. The **range** is the difference between the largest number and the smallest number in a set of data.

Example

The speeds (in m/s) for a race car during five different time trials are 39, 37, 44, 36, and 44.

To find the mean:

Step 1 Find the sum of the numbers.

$$39 + 37 + 44 + 36 + 44 = 200$$

Step 2 Divide the sum by the number of items, which is 5.

$$200 \div 5 = 40$$

The mean is 40 m/s.

To find the median:

Step 1 Arrange the measures from least to greatest.

36, 37, 39, 44, 44

Step 2 Determine the middle measure.

36, 37, <u>39</u>, 44, 44

The median is 39 m/s.

To find the mode:

Step 1 Group the numbers that are the same together.

44, 44, 36, 37, 39

Step 2 Determine the number that occurs most in the set.

<u>44, 44,</u> 36, 37, 39

The mode is 44 m/s.

To find the range:

Step 1 Arrange the measures from greatest to least.

44, 44, 39, 37, 36

Step 2 Determine the greatest and least measures in the set.

<u>44,</u> 44, 39, 37, 36

Step 3 Find the difference between the greatest and least measures.

$$44 - 36 = 8$$

The range is 8 m/s.

Practice Problem Find the mean, median, mode, and range for the data set 8, 4, 12, 8, 11, 14, 16.

A **frequency table** shows how many times each piece of data occurs, usually in a survey. **Table 1** below shows the results of a student survey on favorite color.

Table 1 Student Color Choice		
Color	Tally	Frequency
red	IIII	4
blue	ⅣⅩ	5
black	II	2
green	III	3
purple	ⅣⅩ II	7
yellow	ⅣⅩ I	6

Based on the frequency table data, which color is the favorite?

Use Geometry

The branch of mathematics that deals with the measurement, properties, and relationships of points, lines, angles, surfaces, and solids is called geometry.

Perimeter The **perimeter** (P) is the distance around a geometric figure. To find the perimeter of a rectangle, add the length and width and multiply that sum by two, or $2(l + w)$. To find perimeters of irregular figures, add the length of the sides.

Example 1

Find the perimeter of a rectangle that is 3 m long and 5 m wide.

Step 1 You know that the perimeter is 2 times the sum of the width and length.

$$P = 2(3 \text{ m} + 5 \text{ m})$$

Step 2 Find the sum of the width and length.

$$P = 2(8 \text{ m})$$

Step 3 Multiply by 2.

$$P = 16 \text{ m}$$

The perimeter is 16 m.

Example 2

Find the perimeter of a shape with sides measuring 2 cm, 5 cm, 6 cm, 3 cm.

Step 1 You know that the perimeter is the sum of all the sides.

$$P = 2 + 5 + 6 + 3$$

Step 2 Find the sum of the sides.

$$P = 2 + 5 + 6 + 3$$
$$P = 16$$

The perimeter is 16 cm.

Practice Problem Find the perimeter of a rectangle with a length of 18 m and a width of 7 m.

Practice Problem Find the perimeter of a triangle measuring 1.6 cm by 2.4 cm by 2.4 cm.

Area of a Rectangle The **area** (A) is the number of square units needed to cover a surface. To find the area of a rectangle, multiply the length times the width, or $l \times w$. When finding area, the units also are multiplied. Area is given in square units.

Example

Find the area of a rectangle with a length of 1 cm and a width of 10 cm.

Step 1 You know that the area is the length multiplied by the width.

$$A = (1 \text{ cm} \times 10 \text{ cm})$$

Step 2 Multiply the length by the width. Also multiply the units.

$$A = 10 \text{ cm}^2$$

The area is 10 cm².

Practice Problem Find the area of a square whose sides measure 4 m.

Area of a Triangle To find the area of a triangle, use the formula:

$$A = \frac{1}{2}(\text{base} \times \text{height})$$

The base of a triangle can be any of its sides. The height is the perpendicular distance from a base to the opposite endpoint, or vertex.

Example

Find the area of a triangle with a base of 18 m and a height of 7 m.

Step 1 You know that the area is $\frac{1}{2}$ the base times the height.

$$A = \frac{1}{2}(18 \text{ m} \times 7 \text{ m})$$

Step 2 Multiply $\frac{1}{2}$ by the product of 18×7. Multiply the units.

$$A = \frac{1}{2}(126 \text{ m}^2)$$
$$A = 63 \text{ m}^2$$

The area is 63 m².

Practice Problem Find the area of a triangle with a base of 27 cm and a height of 17 cm.

SCIENCE SKILL HANDBOOK

MATH SKILL HANDBOOK

FOLDABLES HANDBOOK

REFERENCE HANDBOOK

GLOSSARY/ GLOSARIO

INDEX

SCIENCE SKILL HANDBOOK

MATH SKILL HANDBOOK

FOLDABLES HANDBOOK

REFERENCE HANDBOOK

GLOSSARY/ GLOSARIO

INDEX

Circumference of a Circle The **diameter** (d) of a circle is the distance across the circle through its center, and the **radius** (r) is the distance from the center to any point on the circle. The radius is half of the diameter. The distance around the circle is called the **circumference** (C). The formula for finding the circumference is:

$C = 2\pi r$ or $C = \pi d$

The circumference divided by the diameter is always equal to 3.1415926… This nonterminating and nonrepeating number is represented by the Greek letter π (pi). An approximation often used for π is 3.14.

Example 1

Find the circumference of a circle with a radius of 3 m.

Step 1 You know the formula for the circumference is 2 times the radius times π.

$C = 2\pi(3)$

Step 2 Multiply 2 times the radius.

$C = 6\pi$

Step 3 Multiply by π.

$C \approx 19$ m

The circumference is about 19 m.

Example 2

Find the circumference of a circle with a diameter of 24.0 cm.

Step 1 You know the formula for the circumference is the diameter times π.

$C = \pi(24.0)$

Step 2 Multiply the diameter by π.

$C \approx 75.4$ cm

The circumference is about 75.4 cm.

Practice Problem Find the circumference of a circle with a radius of 19 cm.

Area of a Circle The formula for the area of a circle is: $A = \pi r^2$

Example 1

Find the area of a circle with a radius of 4.0 cm.

Step 1 $A = \pi(4.0)^2$

Step 2 Find the square of the radius.

$A = 16\pi$

Step 3 Multiply the square of the radius by π.

$A \approx 50$ cm^2

The area of the circle is about 50 cm^2.

Example 2

Find the area of a circle with a radius of 225 m.

Step 1 $A = \pi(225)^2$

Step 2 Find the square of the radius.

$A = 50625\pi$

Step 3 Multiply the square of the radius by π.

$A \approx 159043.1$

The area of the circle is about 159043.1 m^2.

Example 3

Find the area of a circle whose diameter is 20.0 mm.

Step 1 Remember that the radius is half of the diameter.

$A = \pi\left(\frac{20.0}{2}\right)^2$

Step 2 Find the radius.

$A = \pi(10.0)^2$

Step 3 Find the square of the radius.

$A = 100\pi$

Step 4 Multiply the square of the radius by π.

$A \approx 314$ mm^2

The area of the circle is about 314 mm^2.

Practice Problem Find the area of a circle with a radius of 16 m.

Volume The measure of space occupied by a solid is the **volume** (V). To find the volume of a rectangular solid multiply the length times width times height, or $V = l \times w \times h$. It is measured in cubic units, such as cubic centimeters (cm^3).

Example

Find the volume of a rectangular solid with a length of 2.0 m, a width of 4.0 m, and a height of 3.0 m.

Step 1 You know the formula for volume is the length times the width times the height.

$$V = 2.0 \text{ m} \times 4.0 \text{ m} \times 3.0 \text{ m}$$

Step 2 Multiply the length times the width times the height.

$$V = 24 \text{ m}^3$$

The volume is 24 m^3.

Practice Problem Find the volume of a rectangular solid that is 8 m long, 4 m wide, and 4 m high.

To find the volume of other solids, multiply the area of the base times the height.

Example 1

Find the volume of a solid that has a triangular base with a length of 8.0 m and a height of 7.0 m. The height of the entire solid is 15.0 m.

Step 1 You know that the base is a triangle, and the area of a triangle is $\frac{1}{2}$ the base times the height, and the volume is the area of the base times the height.

$$V = \left[\tfrac{1}{2}(b \times h)\right] \times 15$$

Step 2 Find the area of the base.

$$V = \left[\tfrac{1}{2}(8 \times 7)\right] \times 15$$
$$V = \left(\tfrac{1}{2} \times 56\right) \times 15$$

Step 3 Multiply the area of the base by the height of the solid.

$$V = 28 \times 15$$
$$V = 420 \text{ m}^3$$

The volume is 420 m^3.

Example 2

Find the volume of a cylinder that has a base with a radius of 12.0 cm, and a height of 21.0 cm.

Step 1 You know that the base is a circle, and the area of a circle is the square of the radius times π, and the volume is the area of the base times the height.

$$V = (\pi r^2) \times 21$$
$$V = (\pi 12^2) \times 21$$

Step 2 Find the area of the base.

$$V = 144\pi \times 21$$
$$V = 452 \times 21$$

Step 3 Multiply the area of the base by the height of the solid.

$$V \approx 9{,}500 \text{ cm}^3$$

The volume is about 9,500 cm^3.

Example 3

Find the volume of a cylinder that has a diameter of 15 mm and a height of 4.8 mm.

Step 1 You know that the base is a circle with an area equal to the square of the radius times π. The radius is one-half the diameter. The volume is the area of the base times the height.

$$V = (\pi r^2) \times 4.8$$
$$V = \left[\pi\left(\tfrac{1}{2} \times 15\right)^2\right] \times 4.8$$
$$V = (\pi 7.5^2) \times 4.8$$

Step 2 Find the area of the base.

$$V = 56.25\pi \times 4.8$$
$$V \approx 176.71 \times 4.8$$

Step 3 Multiply the area of the base by the height of the solid.

$$V \approx 848.2$$

The volume is about 848.2 mm^3.

Practice Problem Find the volume of a cylinder with a diameter of 7 cm in the base and a height of 16 cm.

SCIENCE SKILL HANDBOOK

MATH SKILL HANDBOOK

FOLDABLES HANDBOOK

REFERENCE HANDBOOK

GLOSSARY/ GLOSARIO

INDEX

Science Applications

SCIENCE SKILL HANDBOOK

MATH SKILL HANDBOOK

FOLDABLES HANDBOOK

REFERENCE HANDBOOK

GLOSSARY/ GLOSARIO

INDEX

Measure in SI

The metric system of measurement was developed in 1795. A modern form of the metric system, called the International System (SI), was adopted in 1960 and provides the standard measurements that all scientists around the world can understand.

The SI system is convenient because unit sizes vary by powers of 10. Prefixes are used to name units. Look at **Table 2** for some common SI prefixes and their meanings.

Table 2 Common SI Prefixes

Prefix	Symbol	Meaning	
kilo–	k	1,000	thousandth
hecto–	h	100	hundred
deka–	da	10	ten
deci–	d	0.1	tenth
centi–	c	0.01	hundreth
milli–	m	0.001	thousandth

Example

How many grams equal one kilogram?

Step 1   Find the prefix *kilo–* in **Table 2.**

Step 2   Using **Table 2,** determine the meaning of *kilo–*. According to the table, it means 1,000. When the prefix *kilo–* is added to a unit, it means that there are 1,000 of the units in a "kilounit."

Step 3   Apply the prefix to the units in the question. The units in the question are grams. There are 1,000 grams in a kilogram.

Practice Problem Is a milligram larger or smaller than a gram? How many of the smaller units equal one larger unit? What fraction of the larger unit does one smaller unit represent?

Dimensional Analysis

Convert SI Units In science, quantities such as length, mass, and time sometimes are measured using different units. A process called dimensional analysis can be used to change one unit of measure to another. This process involves multiplying your starting quantity and units by one or more conversion factors. A conversion factor is a ratio equal to one and can be made from any two equal quantities with different units. If 1,000 mL equal 1 L then two ratios can be made.

$$\frac{1{,}000 \text{ mL}}{1 \text{ L}} = \frac{1 \text{ L}}{1{,}000 \text{ mL}} = 1$$

One can convert between units in the SI system by using the equivalents in **Table 2** to make conversion factors.

Example

How many cm are in 4 m?

Step 1   Write conversion factors for the units given. From **Table 2,** you know that 100 cm = 1 m. The conversion factors are

$$\frac{100 \text{ cm}}{1 \text{ m}} \text{ and } \frac{1 \text{ m}}{100 \text{ cm}}$$

Step 2   Decide which conversion factor to use. Select the factor that has the units you are converting from (m) in the denominator and the units you are converting to (cm) in the numerator.

$$\frac{100 \text{ cm}}{1 \text{ m}}$$

Step 3   Multiply the starting quantity and units by the conversion factor. Cancel the starting units with the units in the denominator. There are 400 cm in 4 m.

$$4 \text{ m} = \frac{100 \text{ cm}}{1 \text{ m}} = 400 \text{ cm}$$

Practice Problem How many milligrams are in one kilogram? (Hint: You will need to use two conversion factors from **Table 2.**)

Table 3 Unit System Equivalents

Type of Measurement	Equivalent
Length	1 in = 2.54 cm 1 yd = 0.91 m 1 mi = 1.61 km
Mass and weight*	1 oz = 28.35 g 1 lb = 0.45 kg 1 ton (short) = 0.91 tonnes (metric tons) 1 lb = 4.45 N
Volume	$1 \text{ in}^3 = 16.39 \text{ cm}^3$ 1 qt = 0.95 L 1 gal = 3.78 L
Area	$1 \text{ in}^2 = 6.45 \text{ cm}^2$ $1 \text{ yd}^2 = 0.83 \text{ m}^2$ $1 \text{ mi}^2 = 2.59 \text{ km}^2$ 1 acre = 0.40 hectares
Temperature	$°C = \dfrac{(°F - 32)}{1.8}$ $K = °C + 273$

*Weight is measured in standard Earth gravity.

Convert Between Unit Systems Table 3 gives a list of equivalents that can be used to convert between English and SI units.

Example

If a meterstick has a length of 100 cm, how long is the meterstick in inches?

Step 1 Write the conversion factors for the units given. From **Table 3,** 1 in = 2.54 cm.

$$\frac{1 \text{ in}}{2.54 \text{ cm}} \text{ and } \frac{2.54 \text{ cm}}{1 \text{ in}}$$

Step 2 Determine which conversion factor to use. You are converting from cm to in. Use the conversion factor with cm on the bottom.

$$\frac{1 \text{ in}}{2.54 \text{ cm}}$$

Step 3 Multiply the starting quantity and units by the conversion factor. Cancel the starting units with the units in the denominator. Round your answer to the nearest tenth.

$$100 \text{ cm} \times \frac{1 \text{ in}}{2.54 \text{ cm}} = 39.37 \text{ in}$$

The meterstick is about 39.4 in long.

Practice Problem 1 A book has a mass of 5 lb. What is the mass of the book in kg?

Practice Problem 2 Use the equivalent for in and cm (1 in = 2.54 cm) to show how $1 \text{ in}^3 \approx 16.39 \text{ cm}^3$.

SCIENCE SKILL HANDBOOK

MATH SKILL HANDBOOK

FOLDABLES HANDBOOK

REFERENCE HANDBOOK

GLOSSARY/ GLOSARIO

INDEX

SCIENCE SKILL HANDBOOK

MATH SKILL HANDBOOK

FOLDABLES HANDBOOK

REFERENCE HANDBOOK

GLOSSARY/ GLOSARIO

INDEX

Precision and Significant Digits

When you make a measurement, the value you record depends on the precision of the measuring instrument. This precision is represented by the number of significant digits recorded in the measurement. When counting the number of significant digits, all digits are counted except zeros at the end of a number with no decimal point such as 2,050, and zeros at the beginning of a decimal such as 0.03020. When adding or subtracting numbers with different precision, round the answer to the smallest number of decimal places of any number in the sum or difference. When multiplying or dividing, the answer is rounded to the smallest number of significant digits of any number being multiplied or divided.

Example

The lengths 5.28 and 5.2 are measured in meters. Find the sum of these lengths and record your answer using the correct number of significant digits.

Step 1   Find the sum.

	5.28 m	2 digits after the decimal
+	5.2 m	1 digit after the decimal
	10.48 m	

Step 2   Round to one digit after the decimal because the least number of digits after the decimal of the numbers being added is 1.

The sum is 10.5 m.

Practice Problem 1 How many significant digits are in the measurement 7,071,301 m? How many significant digits are in the measurement 0.003010 g?

Practice Problem 2 Multiply 5.28 and 5.2 using the rule for multiplying and dividing. Record the answer using the correct number of significant digits.

Scientific Notation

Many times numbers used in science are very small or very large. Because these numbers are difficult to work with scientists use scientific notation. To write numbers in scientific notation, move the decimal point until only one non-zero digit remains on the left. Then count the number of places you moved the decimal point and use that number as a power of ten. For example, the average distance from the Sun to Mars is 227,800,000,000 m. In scientific notation, this distance is 2.278×10^{11} m. Because you moved the decimal point to the left, the number is a positive power of ten.

The mass of an electron is about 0.000 000 000 000 000 000 000 000 000 000 911 kg. Expressed in scientific notation, this mass is 9.11×10^{-31} kg. Because the decimal point was moved to the right, the number is a negative power of ten.

Example

Earth is 149,600,000 km from the Sun. Express this in scientific notation.

Step 1   Move the decimal point until one non-zero digit remains on the left.

1.496 000 00

Step 2   Count the number of decimal places you have moved. In this case, eight.

Step 2   Show that number as a power of ten, 10^8.

Earth is 1.496×10^8 km from the Sun.

Practice Problem 1 How many significant digits are in 149,600,000 km? How many significant digits are in 1.496×10^8 km?

Practice Problem 2 Parts used in a high performance car must be measured to 7×10^{-6} m. Express this number as a decimal.

Practice Problem 3 A CD is spinning at 539 revolutions per minute. Express this number in scientific notation.

Make and Use Graphs

Data in tables can be displayed in a graph—a visual representation of data. Common graph types include line graphs, bar graphs, and circle graphs.

Line Graph A line graph shows a relationship between two variables that change continuously. The independent variable is changed and is plotted on the x-axis. The dependent variable is observed, and is plotted on the y-axis.

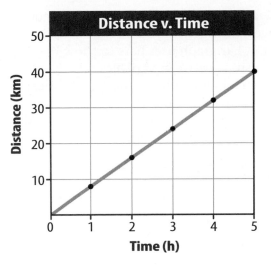

Figure 8 This line graph shows the relationship between distance and time during a bicycle ride.

Example

Draw a line graph of the data below from a cyclist in a long-distance race.

Table 4 Bicycle Race Data	
Time (h)	**Distance (km)**
0	0
1	8
2	16
3	24
4	32
5	40

Step 1 Determine the x-axis and y-axis variables. Time varies independently of distance and is plotted on the x-axis. Distance is dependent on time and is plotted on the y-axis.

Step 2 Determine the scale of each axis. The x-axis data ranges from 0 to 5. The y-axis data ranges from 0 to 50.

Step 3 Using graph paper, draw and label the axes. Include units in the labels.

Step 4 Draw a point at the intersection of the time value on the x-axis and corresponding distance value on the y-axis. Connect the points and label the graph with a title, as shown in **Figure 8**.

Practice Problem A puppy's shoulder height is measured during the first year of her life. The following measurements were collected: (3 mo, 52 cm), (6 mo, 72 cm), (9 mo, 83 cm), (12 mo, 86 cm). Graph this data.

Find a Slope The slope of a straight line is the ratio of the vertical change, rise, to the horizontal change, run.

$$\text{Slope} = \frac{\text{vertical change (rise)}}{\text{horizontal change (run)}} = \frac{\text{change in } y}{\text{change in } x}$$

Example

Find the slope of the graph in **Figure 8**.

Step 1 You know that the slope is the change in y divided by the change in x.

$$\text{Slope} = \frac{\text{change in } y}{\text{change in } x}$$

Step 2 Determine the data points you will be using. For a straight line, choose the two sets of points that are the farthest apart.

$$\text{Slope} = \frac{(40 - 0) \text{ km}}{(5 - 0) \text{ h}}$$

Step 3 Find the change in y and x.

$$\text{Slope} = \frac{40 \text{ km}}{5 \text{ h}}$$

Step 4 Divide the change in y by the change in x.

$$\text{Slope} = \frac{8 \text{ km}}{\text{h}}$$

The slope of the graph is 8 km/h.

SCIENCE SKILL HANDBOOK

MATH SKILL HANDBOOK

FOLDABLES HANDBOOK

REFERENCE HANDBOOK

GLOSSARY/ GLOSARIO

INDEX

SCIENCE SKILL HANDBOOK

MATH SKILL HANDBOOK

FOLDABLES HANDBOOK

REFERENCE HANDBOOK

GLOSSARY/ GLOSARIO

INDEX

Bar Graph To compare data that does not change continuously you might choose a bar graph. A bar graph uses bars to show the relationships between variables. The *x*-axis variable is divided into parts. The parts can be numbers such as years, or a category such as a type of animal. The *y*-axis is a number and increases continuously along the axis.

Example

A recycling center collects 4.0 kg of aluminum on Monday, 1.0 kg on Wednesday, and 2.0 kg on Friday. Create a bar graph of this data.

Step 1 Select the *x*-axis and *y*-axis variables. The measured numbers (the masses of aluminum) should be placed on the *y*-axis. The variable divided into parts (collection days) is placed on the *x*-axis.

Step 2 Create a graph grid like you would for a line graph. Include labels and units.

Step 3 For each measured number, draw a vertical bar above the *x*-axis value up to the *y*-axis value. For the first data point, draw a vertical bar above Monday up to 4.0 kg.

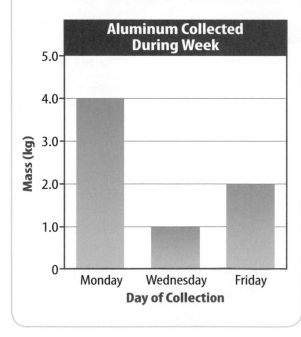

Practice Problem Draw a bar graph of the gases in air: 78% nitrogen, 21% oxygen, 1% other gases.

Circle Graph To display data as parts of a whole, you might use a circle graph. A circle graph is a circle divided into sections that represent the relative size of each piece of data. The entire circle represents 100%, half represents 50%, and so on.

Example

Air is made up of 78% nitrogen, 21% oxygen, and 1% other gases. Display the composition of air in a circle graph.

Step 1 Multiply each percent by 360° and divide by 100 to find the angle of each section in the circle.

$$78\% \times \frac{360°}{100} = 280.8°$$

$$21\% \times \frac{360°}{100} = 75.6°$$

$$1\% \times \frac{360°}{100} = 3.6°$$

Step 2 Use a compass to draw a circle and to mark the center of the circle. Draw a straight line from the center to the edge of the circle.

Step 3 Use a protractor and the angles you calculated to divide the circle into parts. Place the center of the protractor over the center of the circle and line the base of the protractor over the straight line.

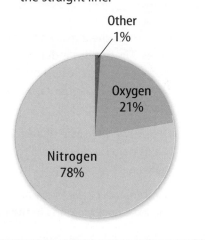

Practice Problem Draw a circle graph to represent the amount of aluminum collected during the week shown in the bar graph to the left.

Student Study Guides & Instructions
By Dinah Zike

1. You will find suggestions for Study Guides, also known as Foldables or books, in each chapter lesson and as a final project. Look at the end of the chapter to determine the project format and glue the Foldables in place as you progress through the chapter lessons.

2. Creating the Foldables or books is simple and easy to do by using copy paper, art paper, and internet printouts. Photocopies of maps, diagrams, or your own illustrations may also be used for some of the Foldables. Notebook paper is the most common source of material for study guides and 83% of all Foldables are created from it. When folded to make books, notebook paper Foldables easily fit into 11″ × 17″ or 12″ × 18″ chapter projects with space left over. Foldables made using photocopy paper are slightly larger and they fit into Projects, but snugly. Use the least amount of glue, tape, and staples needed to assemble the Foldables.

3. Seven of the Foldables can be made using either small or large paper. When 11″ × 17″ or 12″ × 18″ paper is used, these become projects for housing smaller Foldables. Project format boxes are located within the instructions to remind you of this option.

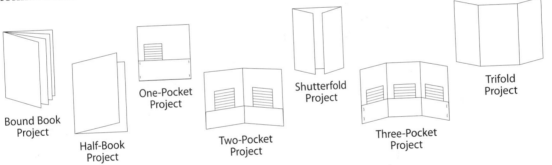

Bound Book Project

Half-Book Project

One-Pocket Project

Two-Pocket Project

Shutterfold Project

Three-Pocket Project

Trifold Project

4. Use one-gallon self-locking plastic bags to store your projects. Place strips of two-inch clear tape along the left, long side of the bag and punch holes through the taped edge. Cut the bottom corners off the bag so it will not hold air. Store this Project Portfolio inside a three-hole binder. To store a large collection of project bags, use a giant laundry-soap box. Holes can be punched in some of the Foldable Projects so they can be stored in a three-hole binder without using a plastic bag. Punch holes in the pocket books before gluing or stapling the pocket.

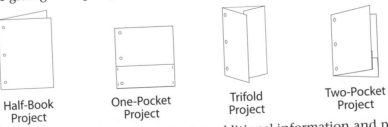

Half-Book Project

One-Pocket Project

Trifold Project

Two-Pocket Project

5. Maximize the use of the projects by collecting additional information and placing it on the back of the project and other unused spaces of the large Foldables.

SCIENCE SKILL HANDBOOK · MATH SKILL HANDBOOK · FOLDABLES HANDBOOK · REFERENCE HANDBOOK · GLOSSARY/GLOSARIO · INDEX

Half-Book Foldable® By Dinah Zike

Step 1 Fold a sheet of notebook or copy paper in half.

Label the exterior tab and use the inside space to write information.

PROJECT FORMAT
Use 11″ × 17″ or 12″ × 18″ paper on the horizontal axis to make a large project book.

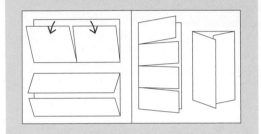

Variations

Paper can be folded horizontally, like a *hamburger* or vertically, like a *hot dog*.

A

B

C Half-books can be folded so that one side is ½ inch longer than the other side. A title or question can be written on the extended tab.

Worksheet Foldable or Folded Book® By Dinah Zike

Step 1 Make a half-book (see above) using work sheets, internet print-outs, diagrams, or maps.

Step 2 Fold it in half again.

Variations

A This folded sheet as a small book with two pages can be used for comparing and contrasting, cause and effect, or other skills.

B When the sheet of paper is open, the four sections can be used separately or used collectively to show sequences or steps.

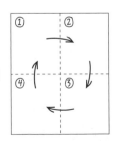

SCIENCE SKILL HANDBOOK

MATH SKILL HANDBOOK

FOLDABLES HANDBOOK

REFERENCE HANDBOOK

GLOSSARY/ GLOSARIO

INDEX

Two-Tab and Concept-Map Foldable® By Dinah Zike

Step 1 Fold a sheet of notebook or copy paper in half vertically or horizontally.

Step 2 Fold it in half again, as shown.

Step 3 Unfold once and cut along the fold line or valley of the top flap to make two flaps.

Variations

A Concept maps can be made by leaving a ½ inch tab at the top when folding the paper in half. Use arrows and labels to relate topics to the primary concept.

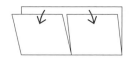

B Use two sheets of paper to make multiple page tab books. Glue or staple books together at the top fold.

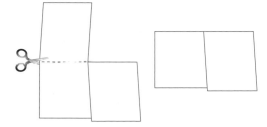

Three-Quarter Foldable® By Dinah Zike

Step 1 Make a two-tab book (see above) and cut the left tab off at the top of the fold line.

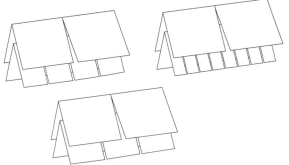

Variations

A Use this book to draw a diagram or a map on the exposed left tab. Write questions about the illustration on the top right tab and provide complete answers on the space under the tab.

B Compose a self-test using multiple choice answers for your questions. Include the correct answer with three wrong responses. The correct answers can be written on the back of the book or upside down on the bottom of the inside page.

SCIENCE SKILL HANDBOOK

MATH SKILL HANDBOOK

FOLDABLES HANDBOOK

REFERENCE HANDBOOK

GLOSSARY/ GLOSARIO

INDEX

SCIENCE SKILL HANDBOOK

MATH SKILL HANDBOOK

FOLDABLES HANDBOOK

REFERENCE HANDBOOK

GLOSSARY/ GLOSARIO

INDEX

Three-Tab Foldable® By Dinah Zike

Step 1 Fold a sheet of paper in half horizontally.

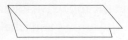

Step 2 Fold into thirds.

Step 3 Unfold and cut along the folds of the top flap to make three sections.

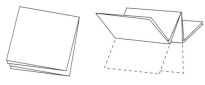

Variations

A Before cutting the three tabs draw a Venn diagram across the front of the book.

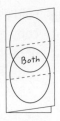

B Make a space to use for titles or concept maps by leaving a ½ inch tab at the top when folding the paper in half.

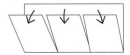

Four-Tab Foldable® By Dinah Zike

Step 1 Fold a sheet of paper in half horizontally.

Step 2 Fold in half and then fold each half as shown below.

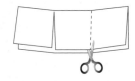

Step 3 Unfold and cut along the fold lines of the top flap to make four tabs.

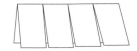

Variations

A Make a space to use for titles or concept maps by leaving a ½ inch tab at the top when folding the paper in half.

B Use the book on the vertical axis, with or without an extended tab.

Folding Fifths for a Foldable® By Dinah Zike

Step 1 Fold a sheet of paper in half horizontally.

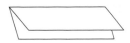

Step 2 Fold again so one-third of the paper is exposed and two-thirds are covered.

Step 3 Fold the two-thirds section in half.

Step 4 Fold the one-third section, a single thickness, backward to make a fold line.

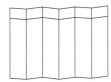

Variations

A Unfold and cut along the fold lines to make five tabs.

B Make a five-tab book with a ½ inch tab at the top (see two-tab instructions).

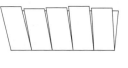

C Use 11″ × 17″ or 12″ × 18″ paper and fold into fifths for a five-column and/or row table or chart.

- -

Folded Table or Chart, and Trifold Foldable® By Dinah Zike

Step 1 Fold a sheet of paper in the required number of vertical columns for the table or chart.

Step 2 Fold the horizontal rows needed for the table or chart.

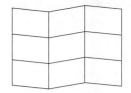

> **PROJECT FORMAT**
> Use 11″ × 17″ or 12″ × 18″ paper and fold it to make a large trifold project book or larger tables and charts.
>
>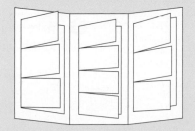

Variations

A Make a trifold by folding the paper into thirds vertically or horizontally.

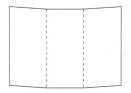

B Make a trifold book. Unfold it and draw a Venn diagram on the inside.

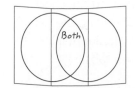

SCIENCE SKILL HANDBOOK
MATH SKILL HANDBOOK
FOLDABLES HANDBOOK
REFERENCE HANDBOOK
GLOSSARY/ GLOSARIO
INDEX

Two or Three-Pockets Foldable® By Dinah Zike

Step 1 Fold up the long side of a horizontal sheet of paper about 5 cm.

Step 2 Fold the paper in half.

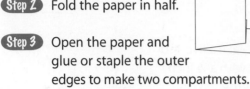

Step 3 Open the paper and glue or staple the outer edges to make two compartments.

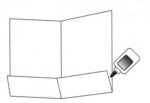

Variations

A Make a multi-page booklet by gluing several pocket books together.

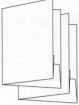

B Make a three-pocket book by using a trifold (see previous instructions).

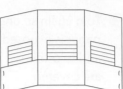

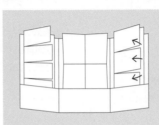

PROJECT FORMAT
Use 11" × 17" or 12" × 18" paper and fold it horizontally to make a large multi-pocket project.

- -

Matchbook Foldable® By Dinah Zike

Step 1 Fold a sheet of paper almost in half and make the back edge about 1–2 cm longer than the front edge.

Step 2 Find the midpoint of the shorter flap.

Step 3 Open the paper and cut the short side along the midpoint making two tabs.

Step 4 Close the book and fold the tab over the short side.

Variations

A Make a single-tab matchbook by skipping Steps 2 and 3.

B Make two smaller matchbooks by cutting the single-tab matchbook in half.

SCIENCE SKILL HANDBOOK

MATH SKILL HANDBOOK

FOLDABLES HANDBOOK

REFERENCE HANDBOOK

GLOSSARY/ GLOSARIO

INDEX

Shutterfold Foldable® By Dinah Zike

Step 1 Begin as if you were folding a vertical sheet of paper in half, but instead of creasing the paper, pinch it to show the midpoint.

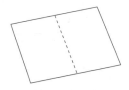

PROJECT FORMAT
Use 11″ × 17″ or 12″ × 18″ paper and fold it to make a large shutterfold project.

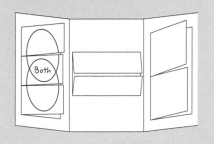

Step 2 Fold the top and bottom to the middle and crease the folds.

Variations

A Use the shutterfold on the horizontal axis.

B Create a center tab by leaving .5–2 cm between the flaps in Step 2.

Four-Door Foldable® By Dinah Zike

Step 1 Make a shutterfold (see above).

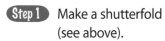

Step 2 Fold the sheet of paper in half.

Step 3 Open the last fold and cut along the inside fold lines to make four tabs.

Variations

A Use the four-door book on the opposite axis.

B Create a center tab by leaving .5–2 cm between the flaps in Step 1.

SCIENCE SKILL HANDBOOK

MATH SKILL HANDBOOK

FOLDABLES HANDBOOK

REFERENCE HANDBOOK

GLOSSARY/ GLOSARIO

INDEX

SCIENCE SKILL HANDBOOK

MATH SKILL HANDBOOK

FOLDABLES HANDBOOK

REFERENCE HANDBOOK

GLOSSARY/ GLOSARIO

INDEX

Bound Book Foldable® By Dinah Zike

Step 1 Fold three sheets of paper in half. Place the papers in a stack, leaving about .5 cm between each top fold. Mark all three sheets about 3 cm from the outer edges.

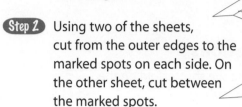

Step 2 Using two of the sheets, cut from the outer edges to the marked spots on each side. On the other sheet, cut between the marked spots.

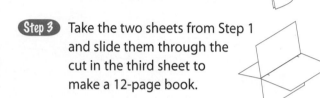

Step 3 Take the two sheets from Step 1 and slide them through the cut in the third sheet to make a 12-page book.

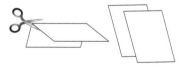

Step 4 Fold the bound pages in half to form a book.

Variation

A Use two sheets of paper to make an eight-page book, or increase the number of pages by using more than three sheets.

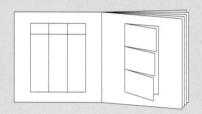

PROJECT FORMAT

Use two or more sheets of 11" × 17" or 12" × 18" paper and fold it to make a large bound book project.

Accordian Foldable® By Dinah Zike

Step 1 Fold the selected paper in half vertically, like a *hamburger*.

Step 2 Cut each sheet of folded paper in half along the fold lines.

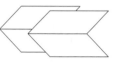

Step 3 Fold each half-sheet almost in half, leaving a 2 cm tab at the top.

Step 4 Fold the top tab over the short side, then fold it in the opposite direction.

Variations

A Glue the straight edge of one paper inside the tab of another sheet. Leave a tab at the end of the book to add more pages.

B Tape the straight edge of one paper to the tab of another sheet, or just tape the straight edges of nonfolded paper end to end to make an accordian.

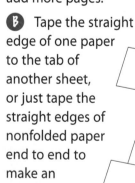

C Use whole sheets of paper to make a large accordian.

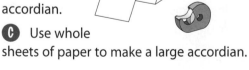

Layered Foldable® By Dinah Zike

Step 1 Stack two sheets of paper about 1–2 cm apart. Keep the right and left edges even.

Step 2 Fold up the bottom edges to form four tabs. Crease the fold to hold the tabs in place.

Step 3 Staple along the folded edge, or open and glue the papers together at the fold line.

Variations

A Rotate the book so the fold is at the top or to the side.

B Extend the book by using more than two sheets of paper.

. . . .

Envelope Foldable® By Dinah Zike

Step 1 Fold a sheet of paper into a *taco*. Cut off the tab at the top.

Step 2 Open the *taco* and fold it the opposite way making another *taco* and an X-fold pattern on the sheet of paper.

Step 3 Cut a map, illustration, or diagram to fit the inside of the envelope.

Step 4 Use the outside tabs for labels and inside tabs for writing information.

Variations

A Use 11″ × 17″ or 12″ × 18″ paper to make a large envelope.

B Cut off the points of the four tabs to make a window in the middle of the book.

SCIENCE SKILL HANDBOOK

MATH SKILL HANDBOOK

FOLDABLES HANDBOOK

REFERENCE HANDBOOK

GLOSSARY/ GLOSARIO

INDEX

Sentence Strip Foldable® By Dinah Zike

Step 1 Fold two sheets of paper in half vertically, like a *hamburger*.

Step 2 Unfold and cut along fold lines making four half sheets.

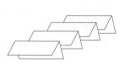

Step 3 Fold each half sheet in half horizontally, like a *hot dog*.

Step 4 Stack folded horizontal sheets evenly and staple together on the left side.

Step 5 Open the top flap of the first sentence strip and make a cut about 2 cm from the stapled edge to the fold line. This forms a flap that can be raisied and lowered. Repeat this step for each sentence strip.

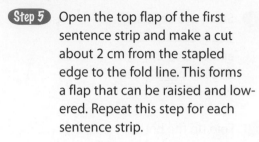

Variations

A Expand this book by using more than two sheets of paper.

B Use whole sheets of paper to make large books.

Pyramid Foldable® By Dinah Zike

Step 1 Fold a sheet of paper into a *taco*. Crease the fold line, but do not cut it off.

Step 2 Open the folded sheet and refold it like a *taco* in the opposite direction to create an X-fold pattern.

Step 3 Cut one fold line as shown, stopping at the center of the X-fold to make a flap.

Step 4 Outline the fold lines of the X-fold. Label the three front sections and use the inside spaces for notes. Use the tab for the title.

Step 5 Glue the tab into a project book or notebook. Use the space under the pyramid for other information.

Title:

Step 6 To display the pyramid, fold the flap under and secure with a paper clip, if needed.

Title:

SCIENCE SKILL HANDBOOK

MATH SKILL HANDBOOK

FOLDABLES HANDBOOK

REFERENCE HANDBOOK

GLOSSARY/ GLOSARIO

INDEX

Single-Pocket or One-Pocket Foldable® By Dinah Zike

Step 1 Using a large piece of paper on a vertical axis, fold the bottom edge of the paper upwards, about 5 cm.

Step 2 Glue or staple the outer edges to make a large pocket.

PROJECT FORMAT
Use 11" × 17" or 12" × 18" paper and fold it vertically or horizontally to make a large pocket project.

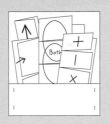

Variations

A Make the one-pocket project using the paper on the horizontal axis.

B To store materials securely inside, fold the top of the paper almost to the center, leaving about 2–4 cm between the paper edges. Slip the Foldables through the opening and under the top and bottom pockets.

Multi-Tab Foldable® By Dinah Zike

Step 1 Fold a sheet of notebook paper in half like a *hot dog*.

Step 2 Open the paper and on one side cut every third line. This makes ten tabs on wide ruled notebook paper and twelve tabs on college ruled.

Step 3 Label the tabs on the front side and use the inside space for definitions or other information.

Variation

A Make a tab for a title by folding the paper so the holes remain uncovered. This allows the notebook Foldable to be stored in a three-hole binder.

SCIENCE SKILL HANDBOOK

MATH SKILL HANDBOOK

FOLDABLES HANDBOOK

REFERENCE HANDBOOK

GLOSSARY/ GLOSARIO

INDEX

SCIENCE SKILL HANDBOOK

MATH SKILL HANDBOOK

FOLDABLES HANDBOOK

REFERENCE HANDBOOK

GLOSSARY/ GLOSARIO

INDEX

PERIODIC TABLE OF THE ELEMENTS

	Element ——— Hydrogen	
	Atomic number ——— 1	
	Symbol ——— **H**	State of matter
	Atomic mass ——— 1.01	

Gas
Liquid
Solid
Synthetic

1

A column in the periodic table is called a **group.**

	1	2	3	4	5	6	7	8	9
1	Hydrogen 1 **H** 1.01								
2	Lithium 3 **Li** 6.94	Beryllium 4 **Be** 9.01							
3	Sodium 11 **Na** 22.99	Magnesium 12 **Mg** 24.31							
4	Potassium 19 **K** 39.10	Calcium 20 **Ca** 40.08	Scandium 21 **Sc** 44.96	Titanium 22 **Ti** 47.87	Vanadium 23 **V** 50.94	Chromium 24 **Cr** 52.00	Manganese 25 **Mn** 54.94	Iron 26 **Fe** 55.85	Cobalt 27 **Co** 58.93
5	Rubidium 37 **Rb** 85.47	Strontium 38 **Sr** 87.62	Yttrium 39 **Y** 88.91	Zirconium 40 **Zr** 91.22	Niobium 41 **Nb** 92.91	Molybdenum 42 **Mo** 95.96	Technetium 43 **Tc** (98)	Ruthenium 44 **Ru** 101.07	Rhodium 45 **Rh** 102.91
6	Cesium 55 **Cs** 132.91	Barium 56 **Ba** 137.33	Lanthanum 57 **La** 138.91	Hafnium 72 **Hf** 178.49	Tantalum 73 **Ta** 180.95	Tungsten 74 **W** 183.84	Rhenium 75 **Re** 186.21	Osmium 76 **Os** 190.23	Iridium 77 **Ir** 192.22
7	Francium 87 **Fr** (223)	Radium 88 **Ra** (226)	Actinium 89 **Ac** (227)	Rutherfordium 104 **Rf** (267)	Dubnium 105 **Db** (268)	Seaborgium 106 **Sg** (271)	Bohrium 107 **Bh** (272)	Hassium 108 **Hs** (270)	Meitnerium 109 **Mt** (276)

The number in parentheses is the mass number of the longest lived isotope for that element.

A row in the periodic table is called a **period.**

Lanthanide series	Cerium 58 **Ce** 140.12	Praseodymium 59 **Pr** 140.91	Neodymium 60 **Nd** 144.24	Promethium 61 **Pm** (145)	Samarium 62 **Sm** 150.36	Europium 63 **Eu** 151.96
Actinide series	Thorium 90 **Th** 232.04	Protactinium 91 **Pa** 231.04	Uranium 92 **U** 238.03	Neptunium 93 **Np** (237)	Plutonium 94 **Pu** (244)	Americium 95 **Am** (243)

Metal

Metalloid

Nonmetal

Recently discovered

			13	**14**	**15**	**16**	**17**	**18**
								Helium 2 **He** 4.00
			Boron 5 **B** 10.81	Carbon 6 **C** 12.01	Nitrogen 7 **N** 14.01	Oxygen 8 **O** 16.00	Fluorine 9 **F** 19.00	Neon 10 **Ne** 20.18

10	**11**	**12**						
			Aluminum 13 **Al** 26.98	Silicon 14 **Si** 28.09	Phosphorus 15 **P** 30.97	Sulfur 16 **S** 32.07	Chlorine 17 **Cl** 35.45	Argon 18 **Ar** 39.95
Nickel 28 **Ni** 58.69	Copper 29 **Cu** 63.55	Zinc 30 **Zn** 65.38	Gallium 31 **Ga** 69.72	Germanium 32 **Ge** 72.64	Arsenic 33 **As** 74.92	Selenium 34 **Se** 78.96	Bromine 35 **Br** 79.90	Krypton 36 **Kr** 83.80
Palladium 46 **Pd** 106.42	Silver 47 **Ag** 107.87	Cadmium 48 **Cd** 112.41	Indium 49 **In** 114.82	Tin 50 **Sn** 118.71	Antimony 51 **Sb** 121.76	Tellurium 52 **Te** 127.60	Iodine 53 **I** 126.90	Xenon 54 **Xe** 131.29
Platinum 78 **Pt** 195.08	Gold 79 **Au** 196.97	Mercury 80 **Hg** 200.59	Thallium 81 **Tl** 204.38	Lead 82 **Pb** 207.20	Bismuth 83 **Bi** 208.98	Polonium 84 **Po** (209)	Astatine 85 **At** (210)	Radon 86 **Rn** (222)
Darmstadtium 110 **Ds** (281)	Roentgenium 111 **Rg** (280)	Copernicium 112 **Cn** (285)	* Ununtrium 113 **Uut** (284)	* Ununquadium 114 **Uuq** (289)	* Ununpentium 115 **Uup** (288)	* Ununhexium 116 **Uuh** (293)		* Ununoctium 118 **Uuo** (294)

* The names and symbols for elements 113–116 and 118 are temporary. Final names will be selected when the elements' discoveries are verified.

Gadolinium 64 **Gd** 157.25	Terbium 65 **Tb** 158.93	Dysprosium 66 **Dy** 162.50	Holmium 67 **Ho** 164.93	Erbium 68 **Er** 167.26	Thulium 69 **Tm** 168.93	Ytterbium 70 **Yb** 173.05	Lutetium 71 **Lu** 174.97
Curium 96 **Cm** (247)	Berkelium 97 **Bk** (247)	Californium 98 **Cf** (251)	Einsteinium 99 **Es** (252)	Fermium 100 **Fm** (257)	Mendelevium 101 **Md** (258)	Nobelium 102 **No** (259)	Lawrencium 103 **Lr** (262)

SCIENCE SKILL HANDBOOK

MATH SKILL HANDBOOK

FOLDABLES HANDBOOK

REFERENCE HANDBOOK

GLOSSARY/ GLOSARIO

INDEX

Topographic Map Symbols

Topographic Map Symbols

Symbol	Description	Symbol	Description
▬▬▬▬▬	Primary highway, hard surface	⌒⌒⌒	Index contour
▬▭▬	Secondary highway, hard surface	·········	Supplementary contour
═══	Light-duty road, hard or improved surface	⌒⌒	Intermediate contour
=========	Unimproved road	⬭	Depression contours
++++++++	Railroad: single track		
╪╪╪╪╪	Railroad: multiple track	▬ ▬ ▬	Boundaries: national
╫╪╫╪╫	Railroads in juxtaposition	▬ ▬ ▬	State
		▬ ▬ ··	County, parish, municipal
▪▖▌▟▓	Buildings	▬ ▬ ▬	Civil township, precinct, town, barrio
♪▴⊞ cem	Schools, church, and cemetery	▬·▪·▪	Incorporated city, village, town, hamlet
▪▭▨▧	Buildings (barn, warehouse, etc.)	▬·▬··	Reservation, national or state
○ ○	Wells other than water (labeled as to type)	----------	Small park, cemetery, airport, etc.
●●●◍	Tanks: oil, water, etc. (labeled only if water)	▬··▬··	Land grant
⊙ ♂	Located or landmark object; windmill	▬▬▬▬	Township or range line, U.S. land survey
⚒ ×	Open pit, mine, or quarry; prospect	--------	Township or range line, approximate location
🟫	Marsh (swamp)		
🟫	Wooded marsh	⬳	Perennial streams
▢	Woods or brushwood	→⟵	Elevated aqueduct
▦	Vineyard	○ ◠	Water well and spring
▨	Land subject to controlled inundation	⌇	Small rapids
▨	Submerged marsh	⌇	Large rapids
▧	Mangrove	⬭	Intermittent lake
▨	Orchard	⌇	Intermittent stream
▢	Scrub	→=====←	Aqueduct tunnel
▦	Urban area	⬭	Glacier
		⌇	Small falls
x7369	Spot elevation	▨	Large falls
670	Water elevation	⬭	Dry lake bed

SIENCE SKILL HANDBOOK

MATH SKILL HANDBOOK

FOLDABLES HANDBOOK

REFERENCE HANDBOOK

GLOSSARY/ GLOSARIO

INDEX

Rocks

Rocks		
Rock Type	**Rock Name**	**Characteristics**
Igneous (intrusive)	Granite	Large mineral grains of quartz, feldspar, hornblende, and mica. Usually light in color.
	Diorite	Large mineral grains of feldspar, hornblende, and mica. Less quartz than granite. Intermediate in color.
	Gabbro	Large mineral grains of feldspar, augite, and olivine. No quartz. Dark in color.
Igneous (extrusive)	Rhyolite	Small mineral grains of quartz, feldspar, hornblende, and mica, or no visible grains. Light in color.
	Andesite	Small mineral grains of feldspar, hornblende, and mica or no visible grains. Intermediate in color.
	Basalt	Small mineral grains of feldspar, augite, and possibly olivine or no visible grains. No quartz. Dark in color.
	Obsidian	Glassy texture. No visible grains. Volcanic glass. Fracture looks like broken glass.
	Pumice	Frothy texture. Floats in water. Usually light in color.
Sedimentary (detrital)	Conglomerate	Coarse grained. Gravel or pebble-size grains.
	Sandstone	Sand-sized grains 1/16 to 2 mm.
	Siltstone	Grains are smaller than sand but larger than clay.
	Shale	Smallest grains. Often dark in color. Usually platy.
Sedimentary (chemical or organic)	Limestone	Major mineral is calcite. Usually forms in oceans and lakes. Often contains fossils.
	Coal	Forms in swampy areas. Compacted layers of organic material, mainly plant remains.
Sedimentary (chemical)	Rock Salt	Commonly forms by the evaporation of seawater.
Metamorphic (foliated)	Gneiss	Banding due to alternate layers of different minerals, of different colors. Parent rock often is granite.
	Schist	Parallel arrangement of sheetlike minerals, mainly micas. Forms from different parent rocks.
	Phyllite	Shiny or silky appearance. May look wrinkled. Common parent rocks are shale and slate.
	Slate	Harder, denser, and shinier than shale. Common parent rock is shale.
Metamorphic (nonfoliated)	Marble	Calcite or dolomite. Common parent rock is limestone.
	Soapstone	Mainly of talc. Soft with greasy feel.
	Quartzite	Hard with interlocking quartz crystals. Common parent rock is sandstone.

SCIENCE SKILL HANDBOOK

MATH SKILL HANDBOOK

FOLDABLES HANDBOOK

REFERENCE HANDBOOK

GLOSSARY/ GLOSARIO

INDEX

Minerals

Mineral (formula)	Color	Streak	Hardness Pattern	Breakage Properties	Uses and Other
Graphite (C)	black to gray	black to gray	1–1.5	basal cleavage (scales)	pencil lead, lubricants for locks, rods to control some small nuclear reactions, battery poles
Galena (PbS)	gray	gray to black	2.5	cubic cleavage perfect	source of lead, used for pipes, shields for X rays, fishing equipment sinkers
Hematite (Fe_2O_3)	black or reddish-brown	reddish-brown	5.5–6.5	irregular fracture	source of iron; converted to pig iron, made into steel
Magnetite (Fe_3O_4)	black	black	6	conchoidal fracture	source of iron, attracts a magnet
Pyrite (FeS_2)	light, brassy, yellow	greenish-black	6–6.5	uneven fracture	fool's gold
Talc ($Mg_3Si_4O_{10}(OH)_2$)	white, greenish	white	1	cleavage in one direction	used for talcum powder, sculptures, paper, and tabletops
Gypsum ($CaSO_4{\cdot}2H_2O$)	colorless, gray, white, brown	white	2	basal cleavage	used in plaster of paris and dry wall for building construction
Sphalerite (ZnS)	brown, reddish-brown, greenish	light to dark brown	3.5–4	cleavage in six directions	main ore of zinc; used in paints, dyes, and medicine
Muscovite ($KAl_3Si_3O_{10}(OH)_2$)	white, light gray, yellow, rose, green	colorless	2–2.5	basal cleavage	occurs in large, flexible plates; used as an insulator in electrical equipment, lubricant
Biotite ($K(Mg,Fe)_3(AlSi_3O_{10})(OH)_2$)	black to dark brown	colorless	2.5–3	basal cleavage	occurs in large, flexible plates
Halite (NaCl)	colorless, red, white, blue	colorless	2.5	cubic cleavage	salt; soluble in water; a preservative

SCIENCE SKILL HANDBOOK
MATH SKILL HANDBOOK
FOLDABLES HANDBOOK
REFERENCE HANDBOOK
GLOSSARY/ GLOSARIO
INDEX

Minerals

Minerals

Mineral (formula)	Color	Streak	Hardness	Breakage Pattern	Uses and Other Properties
Calcite ($CaCO_3$)	colorless, white, pale blue	colorless, white	3	cleavage in three directions	fizzes when HCl is added; used in cements and other building materials
Dolomite ($CaMg(CO_3)_2$)	colorless, white, pink, green, gray, black	white	3.5–4	cleavage in three directions	concrete and cement; used as an ornamental building stone
Fluorite (CaF_2)	colorless, white, blue, green, red, yellow, purple	colorless	4	cleavage in four directions	used in the manufacture of optical equipment; glows under ultraviolet light
Hornblende ($(CaNa)_{2-3}$ $(Mg,Al,$ $Fe)_5-(Al,Si)_2$ Si_6O_{22} $(OH)_2$)	green to black	gray to white	5–6	cleavage in two directions	will transmit light on thin edges; 6-sided cross section
Feldspar ($KAlSi_3O_8$) ($NaAl$ Si_3O_8), ($CaAl_2Si_2$ O_8)	colorless, white to gray, green	colorless	6	two cleavage planes meet at 90° angle	used in the manufacture of ceramics
Augite ((Ca,Na) (Mg,Fe,Al) $(Al,Si)_2 O_6$)	black	colorless	6	cleavage in two directions	square or 8-sided cross section
Olivine ($(Mg,Fe)_2$ SiO_4)	olive, green	none	6.5–7	conchoidal fracture	gemstones, refractory sand
Quartz (SiO_2)	colorless, various colors	none	7	conchoidal fracture	used in glass manufacture, electronic equipment, radios, computers, watches, gemstones

SCIENCE SKILL HANDBOOK

MATH SKILL HANDBOOK

FOLDABLES HANDBOOK

REFERENCE HANDBOOK

GLOSSARY/ GLOSARIO

INDEX

Weather Map Symbols

Sample Station Model

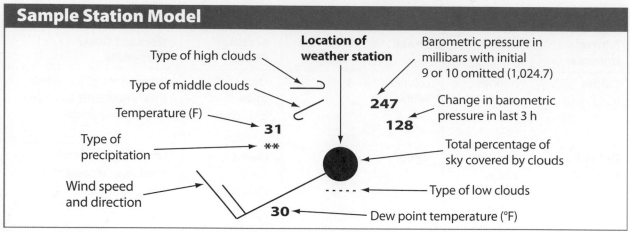

Type of high clouds

Location of weather station

Barometric pressure in millibars with initial 9 or 10 omitted (1,024.7)

Type of middle clouds

247

Change in barometric pressure in last 3 h

Temperature (F)

31

128

Type of precipitation

**

Total percentage of sky covered by clouds

Wind speed and direction

Type of low clouds

30

Dew point temperature (°F)

Sample Plotted Report at Each Station

Precipitation	Wind Speed and Direction	Sky Coverage	Some Types of High Clouds
☰ Fog	◯ 0 calm	◯ No cover	Scattered cirrus
★ Snow	1–2 knots	◐ 1/10 or less	Dense cirrus in patches
● Rain	3–7 knots	◕ 2/10 to 3/10	Veil of cirrus covering entire sky
⊓ Thunderstorm	8–12 knots	◕ 4/10	Cirrus not covering entire sky
' Drizzle	13–17 knots	◖ –	
▽ Showers	18–22 knots	◕ 6/10	
	23–27 knots	◕ 7/10	
	48–52 knots	◕ Overcast with openings	
	1 knot = 1.852 km/h	● Completely overcast	

Some Types of Middle Clouds	Some Types of Low Clouds	Fronts and Pressure Systems	
Thin altostratus layer	⌒ Cumulus of fair weather	Ⓗ or High Ⓛ or Low	Center of high- or low-pressure system
Thick altostratus layer	ᴗ Stratocumulus	▲▲▲▲	Cold front
Thin altostratus in patches	----- Fractocumulus of bad weather	●●●●	Warm front
Thin altostratus in bands	— Stratus of fair weather	●▲●▲	Occluded front
		●▽●▽	Stationary front

Science Skill Handbook

Math Skill Handbook

Foldables Handbook

Reference Handbook

Glossary/ Glosario

Index

Glossary/Glosario

Cómo usar el glosario en español:
1. Busca el término en inglés que desees encontrar.
2. El término en español, junto con la definición, se encuentran en la columna de la derecha.

Pronunciation Key

Use the following key to help you sound out words in the glossary.

a	back (BAK)	ew	food (FEWD)
ay	day (DAY)	yoo	pure (PYOOR)
ah	father (FAH thur)	yew	few (FYEW)
ow	flower (FLOW ur)	uh	comma (CAH muh)
ar	car (CAR)	u (+ con)	rub (RUB)
e	less (LES)	sh	shelf (SHELF)
ee	leaf (LEEF)	ch	nature (NAY chur)
ih	trip (TRIHP)	g	gift (GIHFT)
i (i + com + e)	idea (i DEE uh)	j	gem (JEM)
oh	go (GOH)	ing	sing (SING)
aw	soft (SAWFT)	zh	vision (VIH zhun)
or	orbit (OR buht)	k	cake (KAYK)
oy	coin (COYN)	s	seed, cent (SEED, SENT)
oo	foot (FOOT)	z	zone, raise (ZOHN, RAYZ)

English	**A**	**Español**

absolute age/cast

edad absoluta/contramolde

absolute age: the numerical age, in years, of a rock or object. (p. 345)

edad absoluta: edad numérica, en años, de una roca o de un objeto. (pág. 345)

— B —

basin: area of subsidence; region with low elevation. (p. 279)

cuenca: área de hundimiento; región de elevación baja. (pág. 279)

— C —

carbon film: the fossilized carbon outline of an organism or part of an organism. (p. 122)

película de carbono: contorno de carbono fosilizado de un organismo o parte de un organismo. (pág. 122)

cast: a fossil copy of an organism made when a mold of the organism is filled with sediment or mineral deposits. (p. 341)

contramolde: copia fósil de un organismo producida cuando un molde del organismo se llena con depósitos de sedimento o mineral. (pág. 341)

Science Skill Handbook

Math Skill Handbook

Reference Handbook

Glossary/Glosario

Index

catastrophism: the idea that conditions or creatures on Earth change in quick, violent events. (p. 327)

Cenozoic era: the youngest era of the Phanerozoic eon. (p. 371)

cinder cone: a small, steep-sided volcano that erupts gas-rich, basaltic lava. (p. 310)

coal swamp: an oxygen-poor environment where, over a period of time, decaying plant material changes into coal. (p. 374)

composite volcano: a large, steep-sided volcano that results from explosive eruptions of andesitic and rhyolitic lavas along convergent plate boundaries. (p. 310)

compression: the squeezing force at a convergent boundary. (p. 255)

continental drift: Wegener's hypothesis that suggests that the continents are in constant motion on Earth's surface. (p. 217)

convection: the circulation of particles within a material caused by differences in thermal energy and density. (p. 238)

convergent plate boundary: the boundary between two plates that move toward each other. (p. 235)

correlation: a method used by geologists to fill in the missing gaps in an area's rock record by matching rocks and fossils from separate locations. (p. 340)

catatrofismo: idea de que las condiciones o criaturas en la Tierra cambian mediante eventos rápidos y violentos. (pág. 327)

era Cenozoica: era más joven del eón Fanerozoico. (pág. 371)

cono de ceniza: volcán pequeño de lados empinados que expulsa lava rica en gas basáltico. (pág. 310)

pantano de carbón: medioambiente pobre en oxígeno donde, al paso de un período de tiempo, el material en descomposición de plantas, se transforma en carbón. (pág. 374)

volcán compuesto: volcán grande de lados empinados producido por erupciones explosivas de lavas andesíticas y riolíticas a lo largo de límites convergentes. (pág. 310)

compresión: tensión en un límite convergente. (pág. 255)

deriva continental: hipótesis de Wegener que sugirió que los continentes están en constante movimiento en la superficie de la Tierra. (pág. 217)

convección: circulación de partículas en el interior de un material causada por diferencias en la energía térmica y la densidad. (pág. 238)

límite convergente de placas: límite entre dos placas que se acercan una hacia la otra. (pág. 235)

correlación: método utilizado por los geólogos para completar vacíos en un área de registro de rocas, comparando rocas y fósiles de lugares distanciados. (pág. 340)

D

dinosaur: dominant Mesozoic land vertebrates that walked with their legs positioned directly below their hips. (p. 382)

divergent plate boundary: the boundary between two plates that move away from each other. (p. 235)

dinosaurio: vertebrados dominantes de la tierra del Mesozoico que caminaban con las extremidades ubicadas justo debajo de las caderas. (pág. 382)

límite divergente de placas: límite entre dos placas que se alejan una de la otra. (pág. 235)

E

earthquake: vibrations caused by the rupture and sudden movement of rocks along a break or a crack in Earth's crust. (p. 293)

terremoto: vibraciones causadas por la ruptura y el movimiento repentino de las rocas en una fractura o grieta en la corteza de la Tierra. (pág. 293)

SCIENCE SKILL HANDBOOK

MATH SKILL HANDBOOK

REFERENCE HANDBOOK

GLOSSARY/ GLOSARIO

INDEX

Science Skill Handbook

Math Skill Handbook

Reference Handbook

Glossary/Glosario

Index

eon: the longest unit of geologic time (p. 363)

epicenter: the location on Earth's surface directly above an earthquake's focus. (p. 296)

epoch: a division of geologic time smaller than a period. (p. 363)

era: a large division of geologic time, but smaller than an eon. (p. 363)

eón: unidad más larga del tiempo geológico. (pág. 363)

epicentro: lugar en la superficie de la Tierra justo encima del foco de un terremoto. (pág. 296)

época: división del tiempo geológico más pequeña que un período. (pág. 363)

era: división grande del tiempo geológico, pero más pequeña que un eón. (pág. 363)

F

fault zone: an area of many fractured pieces of crust along a large fault. (p. 265)

fault: a crack or a fracture in Earth's lithosphere along which movement occurs. (p. 295)

fault-block mountain: parallel ridge that forms where blocks of crust move up or down along faults. (p. 272)

focus: a location inside Earth where rocks first move along a fault and from which seismic waves originate. (p. 296)

folded mountain: mountain made of layers of rocks that are folded. (p. 271)

fossil: the preserved remains or evidence of past living organisms. (p. 327)

zona de falla: área de muchos pedazos fracturados de corteza en una falla extensa. (pág. 265)

falla: grieta o fractura en la litosfera de la Tierra en la cual ocurre el movimiento. (pág. 295)

montaña de bloques fallados: dorsal paralela que se forma donde los bloques de corteza se mueven hacia arriba o hacia abajo en las fallas. (pág. 272)

foco: lugar en el interior de la Tierra donde se originan las ondas sísmicas, las cuales son producidas por el movimiento de las rocas a lo largo de un falla. (pág. 296)

montaña plegada: montaña constituida de capas de rocas plegadas. (pág. 271)

fósil: restos conservados o evidencia de organismos vivos del pasado. (pág. 327)

G

geographic isolation: the separation of a population of organisms from the rest of its species due to some physical barrier such as a mountain range or an ocean. (p. 366)

glacial grooves: grooves in solid rock formations made by rocks that are carried by glaciers. (p. 389)

aislamiento geográfico: separación de una población de organismos del resto de su especie debido a alguna barrera física, tal como una cordillera o un océano. (pág. 366)

surcos glaciales: surcos en las formaciones de roca sólida producidos por las rocas transportadas por los glaciares. (pág. 389)

H

half-life: the time required for half of the amount of a radioactive parent element to decay into a stable daughter element. (p. 347)

Holocene epoch: the current epoch of geologic time that began 10,000 years ago. (p. 387)

vida media: tiempo requerido para que la mitad de cierta cantidad de un elemento radiactivo se desintegre en otro elemento estable. (pág. 347)

Holoceno: época actual del tiempo geológico que comenzó hace 10.000 años. (pág. 387)

hot spot: a location where volcanoes form far from plate boundaries. (p. 308)

punto caliente: lugar lejos de los límites de las placas donde se forman volcanes. (pág. 308)

I

ice age: a period of time when a large portion of Earth's surface is covered by glaciers. (p. 389)

inclusion: a piece of an older rock that becomes a part of a new rock. (p. 339)

index fossil: a fossil representative of a species that existed on Earth for a short length of time, was abundant, and inhabited many locations. (p. 341)

inland sea: a body of water formed when ocean water floods continents. (p. 372)

isostasy (i SAHS tuh see): the equilibrium between continental crust and the denser mantle below it. (p. 254)

isotopes: atoms of the same element that have different numbers of neutrons. (p. 346)

era del hielo: período de tiempo cuando los glaciares cubren una gran porción de la superficie de la Tierra. (pág. 389)

inclusión: pedazo de una roca antigua que se convierte en parte de una roca nueva. (pág. 339)

fósil índice: fósil representativo de una especie que existió en la Tierra por un período de tiempo corto, ésta era abundante y habitaba en varios lugares. (pág. 341)

mar interior: cuerpo de agua formado cuando el agua del océano inunda los continentes. (pág. 372)

isostasia: equilibrio entre la corteza continental y el manto más denso debajo de la corteza. (pág. 254)

isótopos: átomos del mismo elemento que tienen números diferentes de neutrones. (pág. 346)

L

land bridge: a landform that connects two continents that were previously separated. (p. 366)

lava: magma that erupts onto Earth's surface. (p. 308)

lithosphere (LIH thuh sfihr): the rigid, outermost layer of Earth that includes the uppermost mantle and crust. (p. 234)

puente terrestre: accidente geográfico que conecta dos continentes que anteriormente estaban separados. (pág. 366)

lava: magma que sale a la superficie de la Tierra. (pág. 308)

litosfera: capa rígida más externa de la Tierra formada por el manto superior y la corteza. (pág. 234)

M

magma: molten rock stored beneath Earth's surface. (p. 307)

magnetic reversal: an event that causes a magnetic field to reverse direction. (p. 228)

mass extinction: the extinction of many species on Earth within a short period of time. (p. 365)

mega-mammal: large mammal of the Cenozoic era. (p. 390)

magma: roca fundida depositada bajo la superficie de la Tierra. (pág. 307)

inversión magnética: evento que causa que un campo magnético invierta su dirección. (pág. 228)

extinción en masa: extinción de muchas especies en la Tierra dentro de un período de tiempo corto. (pág. 365)

mega mamífero: mamífero enorme de la era Cenozoica. (pág. 390)

SCIENCE SKILL HANDBOOK

MATH SKILL HANDBOOK

REFERENCE HANDBOOK

GLOSSARY/ GLOSARIO

INDEX

Mesozoic era: the middle era of the Phanerozoic eon. (p. 371)

mid-ocean ridge: a long, narrow mountain range on the ocean floor; formed by magma at divergent plate boundaries. (p. 225)

mold: the impression of an organism in a rock. (p. 331)

normal polarity: when magnetized objects, such as compass needles, orient themselves to point north. (p. 228)

ocean trench: a deep, underwater trough created by one plate subducting under another plate at a convergent plate boundary. (p. 262)

P

paleontologist: scientist who studies fossils. (p. 332)

Paleozoic era: the oldest era of the Phanerozoic eon. (p. 371)

Pangaea (pan JEE uh): name given to a supercontinent that began to break apart approximately 200 million years ago. (p. 217)

period: a unit of geologic time smaller than an era. (p. 363)

plain: landform with low relief and low elevation. (p. 279)

plate tectonics: theory that Earth's surface is broken into large, rigid pieces that move with respect to each other. (p. 233)

plateau: an area with low relief and high elevation. (p. 280)

Pleistocene epoch: the first epoch of the Quaternary period. (p. 389)

plesiosaur: Mesozoic marine reptile with a small head, long neck, and flippers. (p. 383)

primary wave (also P-wave): a type of seismic wave which causes particles in the ground to move in a push-pull motion similar to a coiled spring. (p. 297)

era Mesozoica: era media del eón Fanerozoico. (pág. 371)

dorsal oceánica: cordillera larga y angosta en el lecho del océano, formada por magma en los límites de las placas divergentes. (pág. 225)

molde: impresión de un organismo en una roca. (pág. 331)

polaridad normal: ocurre cuando los objetos magnetizados, tales como las agujas de la brújula, se orientan a sí mismas para apuntar al norte. (pág. 228)

fosa oceánica: depresión profunda debajo del agua formada por una placa que se desliza debajo de otra placa, en un límite de placas convergentes. (pág. 262)

paleontólogo: científico que estudia los fósiles. (pág. 332)

era Paleozoica: era más antigua del eón Fanerozoico. (pág. 371)

Pangea: nombre dado a un supercontinente que empezó a separarse hace aproximadamente 200 millones de años. (pág. 217)

período: unidad del tiempo geológico más pequeña que una era. (pág. 363)

plano: accidente geográfico de bajo relieve y baja elevación. (pág. 279)

tectónica de placas: teoría que afirma que la superficie de la Tierra está dividida en piezas enormes y rígidas que se mueven una con respecto a la otra. (pág. 233)

meseta: área de bajo relieve y alta elevación. (pág. 280)

época del Pleistoceno: primera época del período Cuaternario. (pág. 389)

plesiosaurio: reptil marino del Mesozoico de cabeza pequeña, cuello largo y aletas. (pág. 383)

onda primaria (también, onda P): tipo de onda sísmica que causa un movimiento de atracción y repulsión en las partículas del suelo, similar a un resorte. (pág. 297)

pterosaur: Mesozoic flying reptile with large, batlike wings. (p. 383)

pterosaurio: reptil volador del Mesozoico de alas grandes parecidas a las del murciélago. (pág. 383)

R

---

radioactive decay: the process by which an unstable element naturally changes into another element that is stable. (p. 346)

relative age: the age of rocks and geologic features compared with other nearby rocks and features. (p. 337)

reversed polarity: when magnetized objects reverse direction and orient themselves to point south. (p. 228)

ridge push: the process that results when magma rises at a mid-ocean ridge and pushes oceanic plates in two different directions away from the ridge. (p. 239)

desintegración radioactiva: proceso poer el cual un elemento inestable cambia naturalmente en otro elemento que es estable. (pág. 346)

edad relativa: edad de las rocas y de las características geológicas comparada con otras rocas cercanas y sus características. (pág. 337)

polaridad inversa: ocurre cuando los objetos magnetizados invierten la dirección y se orientan a sí mismos para apuntar al sur. (pág. 228)

empuje de dorsal: proceso que resulta cuando el magma se levanta en la dorsal oceánica y empuja las placas oceánicas en dos direcciones diferentes, lejos de la dorsal. (pág. 239)

S

---

seafloor spreading: the process by which new oceanic crust forms along a mid-ocean ridge and older oceanic crust moves away from the ridge. (p. 226)

secondary wave (also S-wave): a type of seismic wave that causes particles to move at right angles relative to the direction the wave travels. (p. 297)

seismic wave: energy that travels as vibrations on and in Earth. (p. 296)

seismogram: a graphical illustration of seismic waves. (p. 299)

seismologist (size MAH luh just): scientist that studies earthquakes. (p. 298)

seismometer (size MAH muh ter): an instrument that measures and records ground motion and can be used to determine the distance seismic waves travel. (p. 299)

shear: parallel forces acting in opposite directions at a transform boundary. (p. 255)

expansión del lecho marino: proceso mediante el cual se forma corteza oceánica nueva en la dorsal oceánica, y la corteza oceánica vieja se aleja de la dorsal. (pág. 226)

onda secundaria (también, onda S): tipo de onda sísmica que causa que las partículas se muevan en ángulos rectos respecto a la dirección en que la onda viaja. (pág. 297)

onda sísmica: energía que viaja en forma de vibraciones por encima y dentro de la Tierra. (pág. 296)

sismograma: ilustración gráfica de las ondas sísmicas. (pág. 299)

sismólogo: científico que estudia los terremotos. (pág. 298)

sismómetro: instrumento que mide y registra el movimiento del suelo y que determina la distancia de las ondas sísmicas. (pág. 299)

cizalla: fuerzas paralelas que actúan en direcciones opuestas en un límite transformante. (pág. 255)

shield volcano: a large volcano with gentle slopes of basaltic lavas, common along divergent plate boundaries and oceanic hot spots. (p. 310)

slab pull: the process that results when a dense oceanic plate sinks beneath a more buoyant plate along a subduction zone, pulling the rest of the plate that trails behind it. (p. 239)

strain: a change in the shape of rock caused by stress. (p. 256)

subduction: the process that occurs when one tectonic plate moves under another tectonic plate. (p. 235)

subsidence: the downward vertical motion of Earth's surface. (p. 255)

supercontinent: an ancient landmass that separated into present-day continents. (p. 375)

superposition: the principle that in undisturbed rock layers, the oldest rocks are on the bottom. (p. 338)

surface wave: a type of seismic wave that causes particles in the ground to move up and down in a rolling motion. (p. 297)

volcán escudo: volcán grande con ligeras pendientes de lavas basálticas, común a lo largo de los límites de placas divergentes y puntos calientes oceánicos. (pág. 310)

convergencia de placas: proceso que resulta cuando una placa oceánica densa se hunde debajo de una placa flotante en una zona de subducción, arrastrando el resto de la placa detrás suyo. (pág. 239)

deformación: cambio en la forma de una roca causado por la presión. (pág. 256)

subducción: proceso que ocurre cuando una placa tectónica se mueve debajo de otra placa tectónica. (pág. 235)

hundimiento: movimiento vertical hacia abajo de la superficie de la Tierra. (pág. 255)

supercontinente: antigua masa de tierra que se dividió en los continentes actuales. (pág. 375)

superposición: principio que establece que en las capas de rocas inalteradas, la rocas más viejas se encuentran en la parte inferior. (pág. 338)

onda superficial: tipo de onda sísmica que causa un movimiento de rodamiento hacia arriba y hacia debajo de las partícula en el suelo. (pág. 297)

T

tension: the pulling force at a divergent boundary. (p. 255)

trace fossil: the preserved evidence of the activity of an organism. (p. 331)

transform fault: fault that forms where tectonic plates slide horizontally past each other. (p. 265)

transform plate boundary: the boundary between two plates that slide past each other. (p. 235)

tensión: fuerza de tracción en un límite divergente. (pág. 255)

fósil traza: evidencia conservada de la actividad de un organismo. (pág. 331)

falla transformante: falla que se forma donde las placas tectónicas se deslizan horizontalmente una con respecto a la otra. (pág. 265)

límite de placas transcurrente: límite entre dos placas que se deslizan una con respecto a la otra. (pág. 235)

U

unconformity: a surface where rock has eroded away, producing a break, or gap, in the rock record. (p. 340)

discontinuidad: superficie donde la roca se ha erosionado, produciendo un vacío en el registro geológico sedimentario. (pág. 340)

uniformitarianism: a principle stating that geologic processes that occur today are similar to those that occurred in the past. (p. 328)

uplift: the process that moves large bodies of Earth materials to higher elevations. (p. 255)

uplifted mountain: mountain that forms when large regions rise vertically with very little deformation. (p. 273)

uniformimsmo: principio que establece que los procesos geológicos que ocurren actualmente son similares a aquellos que ocurrieron en el pasado. (pág. 328)

levantamiento: proceso por el cual se mueven grandes cuerpos de materiales de la Tierra hacia elevaciones mayores. (pág. 255)

montaña elevada: montaña que se forma cuando grandes regiones se levantan verticalmente, con muy poca deformación. (pág. 273)

viscosity (vihs KAW sih tee): a measurement of a liquid's resistance to flow. (p. 311)

volcanic arc: a curved line of volcanoes that forms parallel to a plate boundary. (p. 263)

volcanic ash: tiny particles of pulverized volcanic rock and glass. (p. 311)

volcano: a vent in Earth's crust through which molten rock flows. (p. 307)

viscosidad: medida de la resistencia de un líquido a fluir. (pág. 311)

arco volcánico: línea curva de volcanes que se forman paralelos al límite de una placa. (pág. 263)

ceniza volcánica: partículas diminutas de roca y vidrio volcánicos pulverizados. (pág. 311)

volcán: abertura en la corteza terrestre por donde fluye la roca derretida. (pág. 307)

SCIENCE SKILL HANDBOOK

MATH SKILL HANDBOOK

REFERENCE HANDBOOK

GLOSSARY/ GLOSARIO

INDEX

Index

Italic numbers = illustration/photo   **Bold numbers** = vocabulary term
lab = indicates entry is used in a lab on this page

SCIENCE SKILL HANDBOOK

MATH SKILL HANDBOOK

REFERENCE HANDBOOK

GLOSSARY/ GLOSARIO

INDEX

Credits

Photo Credits

Front Cover blickwinkle/Alamy; **Spine-Back Cover** Walter Geiersperger/ CORBIS; **Connect Ed** (t)Richard Hutchings, (c)Getty Images, (b)Jupiter Images/Thinkstock/Alamy; **i** Thinkstock/Getty Images; **iv** Ransom Studios **viii–ix** The McGraw-Hill Companies; **ix** (b)Fancy Photography/ **212** (t to b) Robert Harding/photolibrary.com; (2)CORBIS; (3)maurizio grimaldi/age fotostock; (4)Neal & Molly Jansen/age fotostock; (5)Christine Strover/ Alamy; (6)Photolibrary/age fotostock; **213** (t to b)Jeff Vinnick/Getty Images; (2)Jeff Greenberg/age fotostock; (3)Tom Grill/age fotostock; (4)Neale Clark/ Robert Harding World Imagery/Getty Images; **214–215** Arctic-Images/ Getty Images; **216** Oddur Sigurdsson/Visuals Unlimited, Inc.; **217** Hutchings Photography/Digital Light Source; **219** Walter Geiersperger/CORBIS; **220** Tim Fitzharris/Minden Pictures; **221** Hutchings Photography/Digital Light Source; **223** (l)Peter Johnson/CORBIS; (r)Clare Flemming; **224** Science Source/Photo Researchers; **225** Hutchings Photography/Digital Light Source; **226** Image courtesy of Submarine Ring of Fire 2002 Exploration, NOAA-OE.; **230** Image courtesy of Submarine Ring of Fire 2002 Exploration, NOAA-OE.; **231** (t to b,2,4)Hutchings Photography/Digital Light Source; (r) Dr. Peter Sloss, formerly of NGDC/NOAA/NGDC; (3)Macmillan/McGraw-Hill; **232** NASA; **233** Hutchings Photography/Digital Light Source; **236** (t to b)Dr. Ken MacDonald/Photo Researchers, Inc.; (2)Lloyd Cluff/CORBIS; (3)Jim Richardson/CORBIS; (4)Tony Waltham/Getty Images; **238** Richard Megna/ Fundamental Photographs; **241** (cl)Richard Megna/Fundamental Photographs; **242** (2–5)The McGraw-Hill Companies; (others)Hutchings Photography/Digital Light Source; **250–251** TAO Images Limited/ Photolibrary; **252** Rex A. Stucky/Getty Images; **253** (t)Hutchings Photography/Digital Light Source; (bl)Comstock Images/Alamy; (br)Adam Jones/Getty Images; **254** Ralph A. Clevenger/CORBIS; **256** Hutchings Photography/Digital Light Source; **258** Sinclair Stammers/Photo Researchers, Inc.; **259** (t to b,r,2,4)Hutchings Photography/Digital Light Source; (3,5)The McGraw-Hill Companies; **260** Philippe Bourseiller/Getty Images; **261** (t)Hutchings Photography/Digital Light Source; (b)Tom Till/ Getty Images; **263** D. Falconer/PhotoLink/Getty Images; **264** Geoffrey Morgan/Alamy; **268** Bernhard Edmaier/Photo Researchers, Inc.; **269** Hutchings Photography/Digital Light Source; **271** (t)Tom Bean/CORBIS; (b) DAVID PARKER/SCIENCE PHOTO LIBRARY; **272** Hutchings Photography/ Digital Light Source; **275** (r,4)The Mcgraw-Hill Companies; **276** Digital Vision; **277** Hutchings Photography/Digital Light Source; **279** Courtesy: Jonathan O'Neil/National Science Foundation; **280** C. McIntyre/PhotoLink/ Getty Images; **282** (2,3,6)Macmillan/McGraw-Hill; (others)Hutchings Photography/Digital Light Source; **287** TAO Images Limited/Photolibrary; **290–291** Alberto Garcia/CORBIS; **292** Roger Ressmeyer/CORBIS; **293** (t) Hutchings Photography/Digital Light Source; (b)William S Helsel/Getty Images; **295** Tom Bean/CORBIS; **300** Hutchings Photography/Digital Light Source; **301** (t)Grant Smith/CORBIS; (c)Roger Ressmeyer/CORBIS; (b) Photodisc/Alamy; **304** (t)U.S. Geological Survey; (b)Hutchings Photography/ Digital Light Source; **305** U.S. Geological Survey; **306** Douglas Peebles/ CORBIS; **307** Hutchings Photography/Digital Light Source; **308** (inset) National Oceanic and Atmospheric Administration (NOAA); **309** Corbis Premium RF/Alamy; **310** (tl)Bernd Mellmann/Alamy; (tr)Robert Glusic/ Getty Images; (bl)Michael S. Yamashita/CORBIS; (br)National Geographic/ Getty Images; **311** (t)Digital Vision/Getty Images; (b)David Weintraub/ Photo Researchers, Inc.; **312** (t)Tony Lilley/Alamy; (b)Hutchings Photography/Digital Light Source; **313** (t)Art Wolfe/Getty Images; (b)Image courtesy of Alaska Volcano Observatory/U.S. Geological Survey; **314** (t) Courtesy Chris G. Newhall/U.S. Geological Survey; (b)Science Source/Photo Researchers; **315** (t)Digital Vision/Getty Images; (c)Bernd Mellmann/Alamy; (bl)Robert Glusic/Getty Images; (br)Courtesy Chris G. Newhall/U.S. Geological Survey; **316** (t to b,2)Macmillan/McGraw-Hill; (3)Hutchings Photography/Digital Light Source; (4)Karl Weatherly/CORBIS; **318** Robert

Glusic/Getty Images; **320** NASA/JPL/Cornell; David Weintraub/Photo Researchers, Inc.; **321** Alberto Garcia/CORBIS; **324–325** Corbis/SuperStock; **326** Howard Grey/Getty Images; **327** Hutchings Photography/Digital Light Source; **328** (t)Walt Anderson/Visuals Unlimited, Inc.; (b)Hutchings Photography/Digital Light Source; **329** Wim van Egmond/Visuals Unlimited, Inc.; **330** (t)Staffan Widstrand/CORBIS; (c)David Lyons/Alamy, (b)Scientifica/ Getty Images; **331** (t)Dick Roberts/Visuals Unlimited; (c)Dick Roberts/ Visuals Unlimited; **331** (b)age fotostock/SuperStock; **332** (tl)John Cancalosi/ Alamy; (tr)David Troy/Alam y, (b)Chase Studio Inc.; **333** (t)Jim Zuckerman/ CORBIS; (c)JTB Photo Communications, Inc./Alamy; (b)Jonathan Blair/ CORBIS; **334** (t)Walt Anderson/Visuals Unlimited, Inc.; (c)Scientifica/Getty Images, (b)Chase Studio Inc.; **335** (t,cl)American Museum of Natural History, (b)Wolfgang Kaehler/Alamy; **336** Tom Bean; **337** (t)Hutchings Photography/Digital Light Source; (b)Hemis/CORBIS; **339** Hutchings Photography/Digital Light Source; **340** (t)Ashley Cooper/Alamy; (c)Stephen Reynolds; (b)Marli Miller/Visuals Unlimited, Inc.; **342** Ashley Cooper/Alamy; **343** (t to b)Hutchings Photography/Digital Light Source; (2–4)Macmillan/ McGraw-Hill; **344** Richard T. Nowitz/Photo Researchers, Inc.; **345** (t) Hutchings Photography/Digital Light Source; (b)Stockbyte/PunchStock; **348** Hutchings Photography/Digital Light Source; **352** (l to r, t to b)DK Limited/ CORBIS; (2)Andrew Ward/Life File/Getty Images; (3)John Cancalosi/Alamy; (4,5,12)Kjell B. Sandved/Visuals Unlimited; (6)Andrew Ward/Life File/Getty Images; (7)DK Limited/CORBIS; (8)The Natural History Museum/Alamy; (9) DK Limited/CORBIS; (10)The Natural History Museum/Alamy; (11)Andrew Ward/Life File/Getty Images; (13,14)DK Limited/CORBIS; (15)Tom Bean/ CORBIS; (Coelacanths)DK Limited/CORBIS; (Fossil horseshoe crab)John Cancalosi/Alamy; (crinoid)The Natural History Museum/Alamy; (mid-Cambrian trilobite fossil)Tom Bean/CORBIS; **353** (cw from top)John Cancalosi/Alamy; (2)Andrew Ward/Life File/Getty Images; (3)Kjell B. Sandved/Visuals Unlimited; (4)DK Limited/CORBIS; (5)Tom Bean/CORBIS; (6) DK Limited/CORBIS; (7)The Natural History Museum/Alamy; **354** (t)David Lyons/Alamy; (b)Hemis/CORBIS; **357** Corbis/SuperStock; **360–361** Kevin Schafer/CORBIS; **362** Francois Gohier/Photo Researchers, Inc.; **363** Hutchings Photography/Digital Light Source; **364** (tl)Andy Crawford/Getty Images; (tr,bl)DK Limited/CORBIS; **365** Jonathan Blair/CORBIS; **366** (t) Hutchings Photography/Digital Light Source; (bl)Robert Clay/Alamy; (bc) Panoramic Images/Getty Images; (br)Tom Bean/CORBIS; **367** 2007 Photograph of Ediacara Biota diorama at Smithsonian Institute/Joshua Sherurcij; **368** (t)Andy Crawford/Getty Images; (b)Panoramic Images/Getty Images; **369** (tl)Andrew Ward/Life File/Getty Images; (tc)DK Limited/ CORBIS; (tr)Tom Bean/CORBIS; (bl)John Cancalosi/Alamy; **370** Mark Steinmetz; **371** Hutchings Photography/Digital Light Source; **373** (b)DEA PICTURE LIBRARY/Getty Images; **374** The Field Museum, GEO85637c; **375** Hutchings Photography/Digital Light Source; **376** The Field Museum, GEO85637c; **377** Photodisc/Getty Images; **378** DEA PICTURE LIBRARY/Getty Images; **379** Hutchings Photography/Digital Light Source; **382** Hutchings Photography/Digital Light Source; **383** (t)Naturfoto Honal/CORBIS; (b)Nigel Reed QEDimages/Alamy; **385** (tl)Jin Meng; (tr)American Museum of Natural History; (bl)AMNH/Denis Finnin; (br)American Museum of Natural History; **386** Nik Wheeler/CORBIS; **387** Hutchings Photography/Digital Light Source; **389** Mark Steinmetz; **390** Sinclair Stammers/Photo Researchers, Inc.; **392** (t)Ariadne Van Zandbergen/Lonely Planet Images, Inc.; (b)Getty Images; **393** Ariadne Van Zandbergen/Lonely Planet Images, Inc.; **394** (t to b,3,5,6,7)Hutchings Photography/Digital Light Source; (2,4)Macmillan/ McGraw-Hill; **398** Tom Bean/CORBIS; **399** Kevin Schafer/CORBIS; **SR-00– SR-01** Gallo Images - Neil Overy/Getty Images; **SR-2** Hutchings Photography/Digital Light Source; **SR-6** Michell D. Bridwell/PhotoEdit; **SR-7** (t)The McGraw-Hill Companies, (b)Dominic Oldershaw; **SR-8** StudiOhio; **SR-9** Timothy Fuller; **SR-10** Aaron Haupt; **SR-12** KS Studios; **SR-13 SR-47** Matt Meadows; **SR-48** Stephen Durr, (c)NIBSC/Photo Researchers, Inc., (r) Science VU/Drs. D.T. John & T.B. Cole/Visuals Unlimited, Inc.; **SR-49** (t)Mark

PERIODIC TABLE OF THE ELEMENTS

Element ——— Hydrogen
Atomic number ——— 1
Symbol ——— **H**
Atomic mass ——— 1.01
——— State of matter

🎈 **Gas**
💧 **Liquid**
⬜ **Solid**
⊙ **Synthetic**

A column in the periodic table is called a **group.**

A row in the periodic table is called a **period.**

	1	2	3	4	5	6	7	8	9
1	Hydrogen 1 **H** 🎈 1.01								
2	Lithium 3 **Li** ⬜ 6.94	Beryllium 4 **Be** ⬜ 9.01							
3	Sodium 11 **Na** ⬜ 22.99	Magnesium 12 **Mg** ⬜ 24.31							
4	Potassium 19 **K** ⬜ 39.10	Calcium 20 **Ca** ⬜ 40.08	Scandium 21 **Sc** ⬜ 44.96	Titanium 22 **Ti** ⬜ 47.87	Vanadium 23 **V** ⬜ 50.94	Chromium 24 **Cr** ⬜ 52.00	Manganese 25 **Mn** ⬜ 54.94	Iron 26 **Fe** ⬜ 55.85	Cobalt 27 **Co** ⬜ 58.93
5	Rubidium 37 **Rb** ⬜ 85.47	Strontium 38 **Sr** ⬜ 87.62	Yttrium 39 **Y** ⬜ 88.91	Zirconium 40 **Zr** ⬜ 91.22	Niobium 41 **Nb** ⬜ 92.91	Molybdenum 42 **Mo** ⬜ 95.96	Technetium 43 **Tc** ⊙ (98)	Ruthenium 44 **Ru** ⬜ 101.07	Rhodium 45 **Rh** ⬜ 102.91
6	Cesium 55 **Cs** ⬜ 132.91	Barium 56 **Ba** ⬜ 137.33	Lanthanum 57 **La** ⬜ 138.91	Hafnium 72 **Hf** ⬜ 178.49	Tantalum 73 **Ta** ⬜ 180.95	Tungsten 74 **W** ⬜ 183.84	Rhenium 75 **Re** ⬜ 186.21	Osmium 76 **Os** ⬜ 190.23	Iridium 77 **Ir** ⬜ 192.22
7	Francium 87 **Fr** ⬜ (223)	Radium 88 **Ra** ⬜ (226)	Actinium 89 **Ac** ⬜ (227)	Rutherfordium 104 **Rf** ⊙ (267)	Dubnium 105 **Db** ⊙ (268)	Seaborgium 106 **Sg** ⊙ (271)	Bohrium 107 **Bh** ⊙ (272)	Hassium 108 **Hs** ⊙ (270)	Meitnerium 109 **Mt** ⊙ (276)

The number in parentheses is the mass number of the longest lived isotope for that element.

Lanthanide series	Cerium 58 **Ce** ⬜ 140.12	Praseodymium 59 **Pr** ⬜ 140.91	Neodymium 60 **Nd** ⬜ 144.24	Promethium 61 **Pm** ⊙ (145)	Samarium 62 **Sm** ⬜ 150.36	Europium 63 **Eu** ⬜ 151.96
Actinide series	Thorium 90 **Th** ⬜ 232.04	Protactinium 91 **Pa** ⬜ 231.04	Uranium 92 **U** ⬜ 238.03	Neptunium 93 **Np** ⊙ (237)	Plutonium 94 **Pu** ⊙ (244)	Americium 95 **Am** ⊙ (243)